Springer-Verlag   6900 Heidelberg 1 · Postfach 1780
Telefon (06221) 49101 · Telex 04-61723
1000 Berlin 33 · Heidelberger Platz 3
Telefon (0311) 822001 · Telex 01-83319

Springer-Verlag   New York, NY 10010 · 175, Fifth Avenue
New York Inc.   Telefon 673-2660

**23** Fortschritte der chemischen Forschung
Topics in Current Chemistry

# Molecular Orbitals

Springer-Verlag Berlin Heidelberg GmbH 1971

ISBN 978-3-540-05504-4     ISBN 978-3-540-36866-3 (eBook)
DOI 10.1007/978-3-540-36866-3

© by Springer-Verlag Berlin Heidelberg 1971
Originally published by Springer-Verlag Berlin Heidelberg New York in 1971

# Contents

# MO Theory as a Practical Tool for Studying Chemical Reactivity

**Michael J. S. Dewar**

The University of Texas at Austin, Department of Chemistry, Austin, Texas, (USA)

## Contents

# I. Introduction

The basic problem of chemistry is to interpret chemical reactivity in terms of molecular structure; this then should also be the primary concern of quantum chemistry.

In order to achieve this, we should in principle *calculate the energy of a given aggregate of atoms as a function of their positions in space.* The results can be expressed as a many dimensional potential surface, the minima in which correspond to stable molecules, or aggregates of molecules, while the cols separating the minima correspond to the transition states for reactions leading to their interconversion. If such calculations could be carried out with sufficient accuracy, one could not only

predict the rates of all possible reactions in a given system

but one could also

deduce the detailed geometry of each reaction path,

information which cannot be obtained directly by experiment.

Calculations of this kind, based on rigorous solutions of the Schrödinger equation, are wholly impracticable at present since such solutions can as yet be obtained only for the very simplest systems. Equally, approximate solutions simple enough to be applied to molecules of interest to organic chemists are too inaccurate to give results of chemical value. There is, however, a way out of this impasse, based on analogous situations that have arisen in the past. The accuracy of an approximate treatment can usually be greatly improved by introducing empirical parameters into it; the success of the Debye-Hückel theory of strong electrolytes is a good example. Recent work has shown that an analogous *semiempirical approach,* based on a quantum mechanical treatment simple enough to be applied to molecules of chemical interest, can in fact lead to results of sufficient accuracy and reliability to be of value in the interpretation of reaction mechanisms.

In most cases we can simplify our problem by a familar device. In considering a reaction, $A \rightarrow B$, we can take as a *reaction coordinate* some dimension (e.g. a bond length or angle) that changes during the reaction. If now we minimize the energy of the system for each value of the reaction coordinate, a plot of the resulting energy against the reaction coordinate will give us a section of the potential surface corresponding to the optimum reaction path, i.e. the path following the bottom of the valley in the potential surface leading from the reactants towards the products, over the lowest point of the col separating them, and down the bottom of the valley leading to the products. The minima in the resulting plot then correspond to the reactants and products and the maximum separating them to the transition state. Instead of having to calculate a complete potential surface, we now have only to calculate enough points along the reaction path to locate the maximum in it.

If the results of such calculations are to be chemical value they must be *sufficiently accurate.* We know from both theory and experiment the kind of accuracy required; if rates are to be estimated to a factor of ten, activation enrgies must be accurate to ±1 kcal/mole. Since the methods available to us are crude, they must be tested empirically to see if they achieve this order of accuracy before any reliance can be placed in results obtained from them.

The only points on the potential surface for which experimental data are available are the minima, corresponding to stable molecules whose properties can be studied. The geometry of a molecule corresponds to the coordinates of the corresponding point and its heat of formation to the height of the point in the potential surface. The frequencies of molecular vibrations, determined spectroscopically, allow one to also estimate the curvature of the potential surface at the minimum. It is easily seen that all these quantities must be reproduced by our theoretical treatment if it is to be applied to calculations of reaction paths.

Fig. 1 represents [2] a *reaction coordinate plot* for a system where a reactant $A$ can undergo conversion to one of two products $B$ or $C$. The full line represents the "real" situation; the process $A \rightarrow C$ is favoured since the corresponding transition state $(Y)$ is lower. The other lines represent calculated paths where we suppose the segment $A \rightarrow C$ to be reproduced correctly, but where errors are made in estimating the properties of $B$. An error in estimating its energy corresponds to a vertical displacement in the diagram (– – –), an error in estimating its geometry to a horizontal displacement ($\cdot \cdot \cdot$), and an error in the force constants for stretching or bending of bonds to an error in the

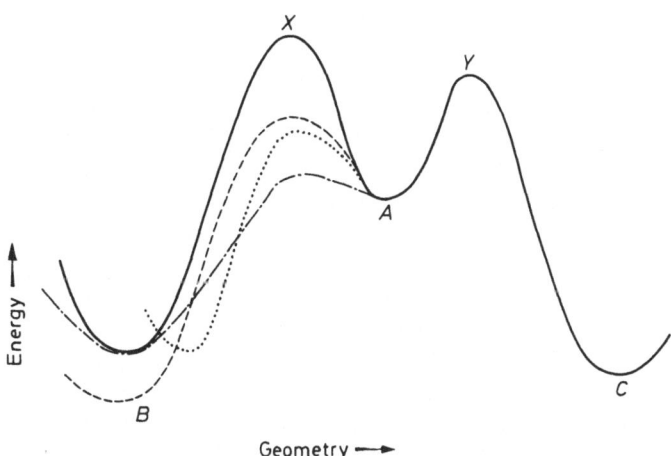

Fig. 1. Effect of errors in calculating properties of a molecule on predictions of the course of a reaction; reproduced from Ref. [2]

curvature at the minimum ($- \cdot - \cdot$). As indicated, any of these errors may lead to a distortion of the surface separating $A$ from $B$, such as to lead to the incorrect prediction that $A \to B$ is favoured over $A \to C$.

We know again from experiment the kind of accuracy needed to avoid such errors; we must be able to estimate energies to the order of $\pm 1$ kcal/mole, bond lengths to the order of $\pm 0.01$ Å, bond angles to the order of $\pm 1°$, and force constants to a few per cent. There is of course no guarantee that a method meeting these conditions will reproduce the intervening parts of the potential surface correctly; we must also test it by calculating activation energies for reactions where experimental values are available. However we can certainly reject any method that fails to meet these conditions since it cannot possibly give satisfactory or reliable results. On this basis we can of course immediately reject any available *"ab initio"* treatment as an *a priori* method for calculating reactivity because the most accurate *ab initio* procedures currently available for organic molecules give energies in error by chemically speaking enormous amounts. Calculations of this kind can be accepted at best on an empirical basis, if, and only if, it can first be shown that some fortuitous cancellation of errors (e.g. of correlation energy in the case of *ab initio* SCF calculations) enables relative energies to be correctly reproduced. This point should be emphasized because numerous calculations of this kind have appeared, and are appearing, in the literature without any such prior justification and have been uncritically accepted by chemists on the misunderstanding that they represent reliable a priori predictions from quantum mechanics. The term *"ab initio"* has been particularly misleading in this connection, giving an erroneous impression of rigour.

A final criterion is provided by *considerations of cost*. If quantum theory is to become a useful chemical tool, it must be possible to carry out calculations for systems of chemical interest at a cost proportionate to the value of the information obtained. In order to predict a reaction path, we must be able to calculate a dozen points along it; moreover we must be able to do this for systems containing at least a dozen atoms other than hydrogen. Each point along the reaction path must be found by minimizing the energy with respect to molecular geometry; since the geometry of a molecule cannot be calculated directly, this can be done only by calculating the energies of a number of different geometries and interpolating. Even the most efficient procedures for doing this require calculations for hundreds of points on the surface; to predict a reaction path therefore requires at least several thousand calculations. If the cost is to be kept within reasonable limits, say 30 000 DM, each individual calculation must take not more than 10 seconds. This at once eliminates ab initio SCF calculations from consideration since the time required is too long by several orders of magnitude. Such a calculation for a molecule the size of decalin would take at least 10 hours on the fastest available digital computers; to calculate a reaction path for a system of this size would cost 10 000 000 DM. Few chemical problems are important enough to justify this kind of expenditure!

Most of the published *ab initio* treatments of reactions have evaded this difficulty by using assumed geometries based on "standard" bond lengths and bond angles, the energy being minimized with respect to variation only of those bonds that change character during the reaction. This unfortunately is an unacceptable simplification since the energy differences we are dealing with are so small. Very small errors in bond lengths and angles, added over a molecule of quite moderate size, can easily amount to 20 kcal/mole or more, comparable in magnitude with the quantity (i.e. the activation energy) that we are trying to calculate. There is of course always a temptation to use approximate treatments on the basis that nothing better is available; this cannot, however, justify calculations based on approximations so extreme as to make the results manifestly meaningless.

## II. Critique of Theoretical Methods

Next we may consider the various quantum mechanical procedures that have been used to calculate potential surfaces for organic reactions.

### A. Ab Initio SCF MO Treatments[a]

These have been based on the Roothaan [3] treatment in which MOs are approximated by linear combinations of basis set functions, usually *Slater-type orbitals* (STOs) or *Gaussian-type orbitals* (GTOs) of the constituent atoms. Technical developments in the evaluation of integrals and the availability of digital computers of the CDC 6600 class have enabled calculations of this kind to be extended to molecules containing six or seven atoms other than hydrogen, while simplified versions, using limited basis sets, have been applied to still larger systems. Since this approach is an approximation to the Hartree-Fock treatment, it is limited to the same ultimate accuracy; the absolute energies of atoms and molecules are therefore in error by the order of 1% of the total binding energy, amounting to thousands of kcal/mole for molecules the size of benzene. As indicated above, calculations of this kind cannot therefore be used in an *a priori* sense for chemical purposes; there is, however, the possibility that they might serve in an empirical sense if it could be shown that the errors cancel. Such could be the case if the total correlation energy of a given set of atoms were independent of their positions in space; heats of atomization of molecules

---

a) For a detailed review, see M. Klessinger, *Fortschritte der Chemischen Forschung,* vol. 9, No. 4, p. 354 (1968).

would then be correctly reproduced and so also the shape of the potential surface for a given aggregate of atoms. Unfortunately this is not so. Calculations of this kind lead to heats of atomization that are in error by chemically speaking *huge amounts,* of the order of 100 kcal/mole per atom. There is therefore a very large change in correlation energy on passing from isolated atoms to molecules.

It is true that recent studies [4] have indicated that this change may be more or less independent of the molecules formed; the relative energies of molecules, or groups of molecules, formed from the same set of atoms are quite well reproduced. This, however, does not extend to species, e.g. radicals, formed by dissociation of molecules; the same total number of bonds must be present if the correlation energy is to remain reasonably constant. Since the intermediate phases of reactions correspond to situations where bonds are partly formed or partly broken, one would not expect such constancy to hold. This could be tested by *calculating reaction paths* for reactions whose activation energies are known experimentally; however few significant calculations of this kind have as yet been reported because, as indicated above, the computation time required to calculate molecular geometries properly is so excessive. In one case where a complete calculation has been reported, i.e. rotation about the double bond in ethylene, the calculated [4b] activation energy (139 kcal/mole) was much greater than that [5] (65 kcal/mole) observed.

If it could be shown that *ab initio* SCF calculations were effective in at least certain connections, they would of course present obvious advantages in that they are based on a rigorous solution of a specific mathematical problem and so involve no parameters. Consequently they can be applied equally well to systems of all kinds, containing any elements. Semiempirical treatments are limited to systems for which parameters have been determined. Even if computation time presents an inseparable barrier to *ab initio* treatments of systems large enough to be of chemical interest, such calculations for simpler systems might prove useful as an aid in developing semiempirical treatments.

## B. Wolfberg-Helmholtz ("Extended Hückel") Method

At the opposite extreme from the *ab initio* SCF methods is the Wolfberg-Helmholtz approximation which Hoffmann [6] has applied extensively to organic problems under the term "extended Hückel method". While this has the advantage of requiring very little computation time, the results are so unreliable that the method is essentially useless for the calculation of potential surfaces. Not only are the errors in heats of atomization comparable with those given by *ab initio* SCF but they are not even the same for isomers. A good example is provided by cyclopropanone *(1)* which is predicted [7] to be less stable than the isomeric zwitterion *2,* a result at variance with the available evidence [8] concerning the mechanism of the Favorskii reaction. Bond lengths and force constants are also subject to large errors; for example the equilibrium bond length in $H_2$ is

predicted to be zero! Calculations of reaction paths by this procedure cannot be taken seriously even in a qualitative sense.

## C. CNDO/2

This semiempirical treatment (CNDO ≡ *Complete Neglect of Differential Overlap*), introduced by Pople *et al.* [9], is derived from the full Roothaan [3] LCAO SCF MO treatment by making the following approximations[a]:

(1) Only valence shell electrons are calculated, these being assumed to move in a fixed core composed of the nuclei and inner shell electrons.

(2) A minimum basis set of *Slater-type orbitals* (STO) is used.

(3) All integrals involving differential overlap between AOs are neglected other than the one-electron resonance integrals.

(4) The two-center repulsion integrals between a pair of centers are assumed to have a common value, i.e. that calculated for the corresponding s–s interaction.

(5) Core-electron attraction integrals are calculated using the Goeppert-Mayer-Sklar approximation with neglect of penetration integrals.

(6) The one-center integrals are determined empirically from spectroscopic data for atoms.

(7) The one-electron resonance integral $\beta_{ij}$ between AOs $\phi_i$ of atom $A$ and $\phi_j$ of atom $B$ is given by:

$$\beta_{ij} = (C_A + C_B) S_{ij} \tag{1}$$

where $S_{ij}$ is the corresponding overlap integral and $C_A$, $C_B$ are empirical constants characteristic of the atoms and chosen to make the results for simple molecules correspond as closely as possible to those given by initio SCF calculations.

Since this treatment is parametrized to mimic the results of *ab initio* calculations, it is not surprising to find that it gives equally inaccurate estimates of heats of atomization. The errors in bond lengths and bond angles are also greater, while force constants are in error by a factor of two or three. While the computation time required is much less than for the ab initio methods,

---

a) For a detailed review of this and the other semiempirical methods discussed below, see G. Klopman and B. O'Leary, *Fortschritte der Chemischen Forschung*, vol. 15, No. 4, p. 447 (1970).

being within reach of our target, the errors in heats of atomization are no longer constant for isomers so that heats of reaction are not correctly predicted. The few applications to reactions, in which geometries have been correctly calculated, have led to estimates of activation energies which are in error by rather large amounts[10]. One must conclude that CNDO/2 does not provide a satisfactory procedure for calculating reaction paths quantitatively. However the method is clearly superior to "extended Hückel" and requires little more computing time; there is therefore no justification of any kind for further use of "extended Hückel" in this connection.

One might add that the failure of CNDO/2 is probably mainly due to the method of parametrization. If a semiempirical method is to be used to estimate heats of formation and molecular geometries, the parameters in it should be chosen accordingly rather than to mimic the results of an approximation known to give unsatisfactory estimates of energies. Recent studies suggest that CNDO/2 may in fact prove useful if properly parametrized. [11]

## D. INDO, PNDO, NDDO

These represent additional approximations intermediate between CNDO/2 and the full Roothaan method. None of them has as yet been used to calculate potential surfaces. INDO [12] ($\equiv$ *Intermediate Neglect of Differential Overlap*) differs from CNDO/2 only in the inclusion of one-center exchange integrals $(ij, ij)$. In NDDO [9] ($\equiv$ *Neglect of Diatomic Differential Overlap*) all integrals involving one-center differential overlap are included; so far very few calculations of any kind have been reported using this approximation. PNDO [13] ($\equiv$ *Partial Neglect of Differential Overlap*) is an approximation intermediate between INDO and NDDO in which the parameters are chosen to reproduce heats of atomization using assumed geometries; since this treatment in its present form fails to give correct estimates of geometries, it cannot be used to calculate potential surfaces.

## E. MINDO/1 and MINDO/2

As indicated above, early attempts to use semiempirical methods had proved unsatisfactory, due to the wrong choice of parameters. A similar situation had existed in the Pople [14] treatment of conjugated molecules using the Hückel $\sigma$, $\pi$ approximation; the parameters in this were chosen to fit spectroscopic data and with these the method gave poor estimates of ground state properties. Subsequent work in our laboratories has shown [15] that this approach can lead to estimates of heats of atomization and molecular geometries that are in almost perfect agreement with experiment if the parameters are chosen to reproduce these quantities.

The success of this $\pi$ approximation led us to attack the problem of three-dimensional molecules in an analogous manner, using one of the simplified SCF MO treatments indicated above. Here of course it is much more difficult to fit heats of atomization and molecular geometries simultaneously since there is no longer a skeleton of localized $\sigma$ bonds to hold the atoms in place; in preliminary studies, using the PNDO or INDO approximations, we used assumed geometries, based on "standard" bond lengths ans bond angles, the parameters being chosen to give the best fit for heats of atomization. Both methods proved encouragingly successful; that based on INDO was termed MINDO [16] (Modified INDO). However neither method gave correct estimates of molecular geometries so neither could be used to calculate potential surfaces.

The choice of parameters in a treatment of this kind presents unexpected problems. The parameters are so interrelated that it is impossible to tell from intuition what will happen if a given parameter is changed. Attempts to get MINDO to give good geometries failed until we devised [2] a computer program to optimize parameters by a least squares method. Using this we were able to develop a new approximation (MINDO/2 [2,17]) which for the first time gave heats of atomization and molecular geometries with an accuracy within sight of that required.

In MINDO/2, the one-center integrals are, as usual, determined from spectroscopic data, essentially in the same way as in INDO. The two-center repulsion integrals $(ii, jj)$ are given by the *Ohno-Klopman* [18] approximation;

$$(ii, jj) = e^2 (r_{ij}^2 + (\rho_i + \rho_j)^2)^{-\frac{1}{2}} \qquad (2)$$

where

$$\rho_i = \tfrac{1}{2} e^2 (ii, ii)^{-1}; \ \rho_j = \tfrac{1}{2} e^2 (jj, jj)^{-1} \qquad (3)$$

The one-electron core resonance integrals $\beta_{ij}^c$ are given by the *Mulliken approximation*;

$$\beta_{ij} = B_{MN} S_{ij} (I_i + I_j) \qquad (4)$$

where $I_i$ and $I_j$ are the valence state ionization potentials of AOs $i$ of atom $M$ and $j$ of atom $N$, $S_{ij}$ is the corresponding overlap integral, and $B_{MN}$ is a parameter characteristic of the atom pair $MN$. The electron-nuclear attraction integrals are set equal to *minus* the corresponding electron-electron repulsions, as in CNDO/2. Finally the core repulsion $(CR)$ is no longer given, as in CNDO/2, by the point charge repulsion $(PCR)$;

$$PCR = \frac{z_M z_N e^2}{r_{MN}} \qquad (5)$$

where $z_M$ and $z_N$ are the formal charges (in units of $e$) on the cores of atoms

9

$M$ and $N$ and $r_{MN}$ the internuclear separation. Instead, we set $CR$ equal to the following parametric function:

$$CR = ER + (PCR - ER)\, e^{\alpha_{MN} r_{MN}} \qquad (6)$$

where $\alpha_{MN}$ is another parameter characteristic of the atom-pair $MN$ and $ER$ is the repulsion between $z_M$ valence electrons on atom $M$ and $z_N$ on atom $N$, i.e.

$$CR = z_M z_N (ii, jj) \qquad (7)$$

$CR$ is thus the total interelectronic repulsion between atoms $M$ and $N$ when each has enough valence electrons to make it neutral.

There are then two parameters per atom pair, $B_{MN}$ and $\alpha_{MN}$; these are found by fitting the heats of atomization of a set of standard molecules and the length of one bond in each. For a set of $n$ different atoms, these are $n(n + 1)$ parameters; values are at present available for the combinations CHON, CHF, and CHCl.

This approach gives quite good estimates of *heats of atomization* and bond lengths; some examples are shown in Table 1. For convience, the calculated heats of atomization have been converted to heats of formation using experimental values for the heats of atomization of elements. As Table 2 shows, the method also gives surprisingly good estimates of force constants; it certainly comes far closer to satisfying the requirements indicated earlier than any other treatment yet proposed. Although not relevant in the present connection, one might add that it also gives good estimates of other ground state properties as well. Dipole moments are reproduced to ±10% and ionization potentials mostly to a few tenths of an electronvolt.

It should be emphasized that the method is still far from perfect. It underestimates strain energies, particularly in four-membered rings, where the error can be 15–20 kcal/mole, and it also gives unsatisfactory results for compounds containing adjacent atoms with lone pairs (e.g. peroxides and hydrazines). Bonds to hydrogen are systematically too long by 0.1 Å, the values in Table 1 being corrected accordingly, and the errors in energies, while much less than those given by other methods are still too large. At the same time the choice of parameters in the present version (i.e. MINDO/2) is certainly not yet optimized. We have already been able to correct the systematic error in bond lengths to hydrogen and we feel confident that the errors in strain energies will soon also be corrected. The inability to deal with lone pairs is, however, inherent in the approximation used here; a treatment in which one-center overlap is neglected is incapable [17] of accounting for the dipolar field of lone pair electrons in hybrid AOs. This difficulty could be avoided by using an analogous version of NDDO, parametrized to reproduce ground state properties. No such treatment has as yet been reported and our own attempts in this direction are still at an interim stage.

Table 1. *Calculated (MINDO/2) and observed Geometries and Heats of Formation*

| Compound and bond lengths (Å) calcd.[a](obsd.[b]) | Heat of formation at 25 °C (kcal/mole) | |
|---|---|---|
| | calcd. | obsd.[b] |
| H,H₂C—CH,H₂ — 1.103(1.093) top, 1.524(1.534) bottom | −21.7 | −20.2 |
| H₂C=CH₂ — 1.093(1.083), 1.337(1.338) | 16.4 | 12.5 |
| H–C≡C–H — 1.069(1.059), 1.206(1.206) | 53.4 | 54.3 |
| H₂C=CH—CH=CH₂ — 1.455(1.467), 1.329(1.343) | 30.9 | 26.3 |
| cyclohexane 1.524(1.534) | −31.3 | −29.4 |
| (CH₃)₂N—CH₂−H — 1.453(1.472) 1.117(1.09, assumed) | 3.3 | − 5.2 |
| CH₃−O−CH=O — 1.224; 1.392 1.334(1.200); (1.437)(1.334) | −84.5 | −81.0 |
| H—ONO₂ — 0.979 (average) (0.980)1.206(1.22) | −36.4 | −32.2 |
| O⋯N(=O)(O) — 1.239(1.241) | −95.8 | −89±5 |

Table 1. (continued)

| Compound and bond lenths (Å) calcd.[a](obsd.[b]) | Heat of formation at 25 °C (kcal/mole) | |
|---|---|---|
| | calcd. | obsd.[b] |
| O⋯N⋯O<br>1.173(1.189) | −10.9 | − 8.1 |
| ←1.355 (1.340)<br>1.473 (1.476) | 56.7 | 69.5 |
| 0.974(0.960)<br>$CH_3$—O—H<br>1.374(1.428) | −53.3 | −48.1 |
| H<br>  C=C=O<br>H<br>1.308  1.189<br>1.091 (1.304) (1.161)<br>(1.083) | −20.4 | −14.6 |
| O=C=O<br>1.162(1.189) | −94.0 | −91.4 |
| O=C=C=C=O<br>1.278(1.294) 1.187(1.168) | −41.9 | −23,4;−47.4 |
| $HCO^+$ | 202.4 | 207 |
| Bicyclo [2,2,2] octane | −25.4 | −31.7[c] |
| Adamantane | −34.8 | −33.9[c] |
| Congressane | −41.7 | −38.3[c] |

a) Bond lengths to hydrogen corrected for systematic error of 0.1 Å; see Ref. [2]
b) For references, see Ref. 17; M.J.S. Dewar, A. Harget, and E. Haselbach, *J. Amer. Chem. Soc., 91,* 7521 (1969).
c) Estimated from the bond energy scheme of J.D. Cox, *Tetrahedron 19,* 1175 (1963).

Table 2. *Calculated (MINDO/2) and Observed Force Constants*

| Compound | Bond | Force constant (mdyne/Å) | |
|---|---|---|---|
| | | Calcd. | Obsd. |
| $CH_4$ | CH | 5.8 | 5.0 |
| $C_2H_6$ | CC | 4.9 | 4.5 |
| | CH | 5.7 | 4.8 |
| $C_2H_4$ | CC | 9.3 | 9.6 |
| | CH | 5.7 | 5.1 |
| $C_2H_2$ | CC | 15.1 | 15.8 |
| | CH | 6.1 | 5.9 |
| $H_2O$ | HO | 10.1 | 7.8 |
| $CH_2O$ | CO | 16.5 | 12.3 |
| $CO_2$ | CO | 22.7 | 16.8 |
| $NH_3$ | NH | 7.7 | 6.4 |
| $HNO_2$ | NO | 8.4 | 7.4 |
| $N_2O$ | NO | 14.4 | 11.4 |

## III. Calculation of Molecular Geometries

A major problem in applying MO theory to organic chemistry is that of calculating molecular geometries. The iterative procedure used in the Westheimer-Allinger [19] approach is not applicable here since too much computation
would be involved; in their treatment an inefficient search procedure can be
used to find the potential minimum since the calculations of energies of individual configurations from empirical potential functions are trivial. Until recently, SCF calculations of geometries had relied on trial-and-error methods.

A *computer program* [20] has now been devised which is economical enough
to make the automatic calculation of geometries feasible. This uses a standard
minimization procedure (Simplex [21]), based on the properties of regular poly-
hedra in $n$-dimensional space. The minimum polyhedron is one with $(n + 1)$
vertices whose faces are equilateral triangles; e.g. the regular tetrahedron in
three dimensions. If we reflect one vertex $(A)$ of the polyhedron in the center
of gravity $(G)$ of the remaining vertices, the resulting polyhedron is a mirror
image of the first, displaced in the direction $AG$. This is illustrated in Fig. 2
for the three-dimensional case.

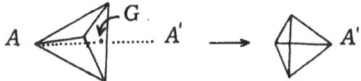

Fig. 2. Inversion of a tetrahedron by reflection of one vertex *(A)* in the centroid *(G)* of the
remaining vertices

13

Any geometry of a molecule can be represented by a point in $n$-dimensional space, $n$ being the number of coordinates (bond lengths, bond angles, dihedral angles, etc.) required to specify it. We pick a point $(A)$ representing our guess at the geometry, construct a sphere of appropriately chosen radius $r$ about $A$, and inscribe a minimum regular polygon in the sphere. The energy corresponding to each of the $(n + 1)$ vertices is then calculated. The point $(B)$ of maximum energy is presumably the one furthest from that $(X)$ representing the true equilibrium geometry; we therefore replace $B$ by its reflection $(C)$ in the centroid of the remaining vertices, thus generating a new polyhedron displaced in the required direction. The energy of $C$ is calculated and the point of highest energy again replaced by its reflection in the centroid of the remaining points. The process is repeated until tests show that the polyhedron contains $X$. We now shrink the polyhedron to some fraction of its original size and continue the treatment. Eventually $X$ will be contained in a polyhedron small enough for the possible error in identifying $X$ with its centroid to be negligible. In its present preliminary form, the calculation converges after about $n^2$ individual SCF calculations. This enables up to twenty variables to be minimized without excessive computation, given that the time required for a molecule with six atoms other than hydrogen is ca 10 seconds on our computer (CDC 6600).

# IV. Applications of MINDO/2 to Some Chemical Problems

Since MINDO/2 seemed to give reasonable estimates of ground state properties, the next step was to study its application to chemical reactions. The rest of this paper describes the results so far obtained.

## A. Barriers to Rotation about C=C Bonds

Calculations of barrier heights for mutual rotation of the terminal methylene groups in ethylene and the cumulenes were reported in the original MINDO/2 paper [2]; since the parameters were subsequently [17] modified somewhat, we have repeated [22] these calculations and are extending them to other olefines.

Some results are shown in Table 3 together with experimental values where available and also some barrier heights calculated by Pople et al [4c] using an *ab initio* SCF method.

Table 3. *Barriers to Rotation About C=C Bonds*

| Compound | Barrier to rotation (kcal/mole) | | |
|---|---|---|---|
| | calcd. (MINDO/2) | obsd. | calcd. (ab initio SCF[4c]) |
| $CH_2=CH_2$ | 54 | 65[a] | 139 |
| $CH_2=C=CH_2$ | 37 | – | 92 |
| $CH_2=C=C=CH_2$ | 32 | 30[a] | 74 |
| $CH2=C=C=C=CH_2$ | 25 | – | – |
| $CH_2=C=C=C=C=CH_2$ | 22 | 20[a] | – |
| (zigzag triene structure) | 47 | 43[b] | |
| (cyclopropylidenecyclopropane structure) | 35 | | |
| (methylenecyclopropene structure) | 31 | | |
| (cyclopropenylidenecyclopropene structure) | 43.4 | | |

a) For references, see Ref. [2]
b) Personal communication from Professor W. von E. Doering.

It will be seen that our results are in good agreement with experiment whereas the *ab initio* values are clearly much too large. Note the low predicted values for methylenecyclopropene, cyclopropylidenecyclopropane, and cyclopropenylidenecyclopropene; it will be interesting to see if these predictions can be confirmed experimentally.

## B. Conformational Isomerisations

Fig. 3 summarizes calculations [23] for the conversion of the chair form of *cyclohexane* to the boat. The calculated difference in energy between the two isomers (5.4 kcal/mole) agrees well with experiment (5.3 kcal/mole). We have not as yet calculated the complete potential surface for the interconversion; however energies have been calculated for three possible transition states (*a – c*) in which four, five, and six carbon atoms respectively are coplanar. In each case the energy was minimized subject to this one constraint. It will be seen that the predicted intermediate is *a,* in agreement with calculations by

15

Hendrickson *et al.* [24] using the Westheimer-Allinger [19] method. Our calculated activation energy (6.0 kcal/mole) is less than those reported by Anet *et al* [25] (10.3, 10.8 kcal/mole). Since, however, the barrier to inversion in perfluorocyclohexane is 7.5 kcal/mole [26], and since the rates of inversion of cyclohexene and perfluorocyclohexane are very similar, the barrier reported for cyclohexane may be too large.

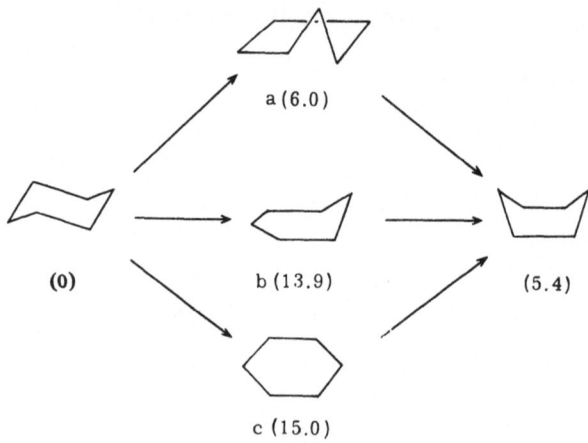

Fig. 3. Chair and boat forms of cyclohexane and possible transition states for their interconversion; energies (kcal/mole) relative to the chair form in parentheses

Fig. 4 summarizes analogous calculations [27] for tub→tub inversion and bond exchange in *cyclooctatetraene (3)*. Inversion must take place via the planar form; we predict the bonds in this to alternate. Inversion should therefore take place without bond exchange; Anet *et al.* [28a] and Roberts *et al.* [28b] have shown this to be the case. Our calculated activation energy (17 kcal/mole) is in reasonable agreement with the observed [28] free energy of activation (13––15 kcal/mole). On the other hand the calculated difference in energy (14 kcal/mole) between the symmetrical planar form *(4)*, with equal bond lengths, and that *(5)* with alternating bonds, is much greater than the difference (~ 2 kcal/mole) between the observed [28] free energies of activation for inversion and bond exchange. We feel this difference is sufficient to exclude the symmetri-

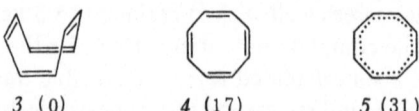

3 (0)        4 (17)        5 (31)

Fig. 4. Calculated energies (kcal/mole) of planar forms of cyclooctatetraene with alternating and equal bond lengths relative to the tub form

cal planar structure as an intermediate in the latter process. Indeed, further calculations [27] indicated that bond exchange can occur as readily in nonplanar forms of cyclooctatetraene as in planar ones; it therefore seems likely that inversion and bond exchange are independent processes.

### C. Cope Rearrangement

The Cope rearrangement of *biallyl (6)* is an intramolecular process that presumably involves the symmetrical intermediate *(7)* as the transition state. Doering and Roth [29] pointed out that 7 could exist in two possible conformations 8 and 9, analogous to the boat and chair forms of cyclohexane, and devised an extremely ingenious experiment to distinguish between them. The reaction is a particularly interesting one since while both paths are "allowed" by arguments based on orbital symmetry [30], that proceeding via the "chair" transition state 9 should be favoured according to arguments based on the aromaticity of the transition state [31].

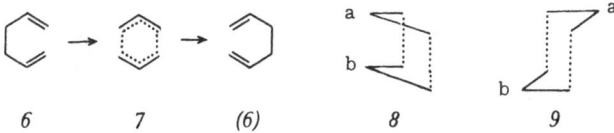

We have calculated [32] the energies of 6, 8 and 9, minimizing the energy with respect to all bond lengths, bond angles and dihedral angles. The only assumption was that 8 and 9 each have a plane of symmetry passing through the central atoms of the two allyl moieties. Fig. 5 shows the calculated geometries and energies relative to 6.

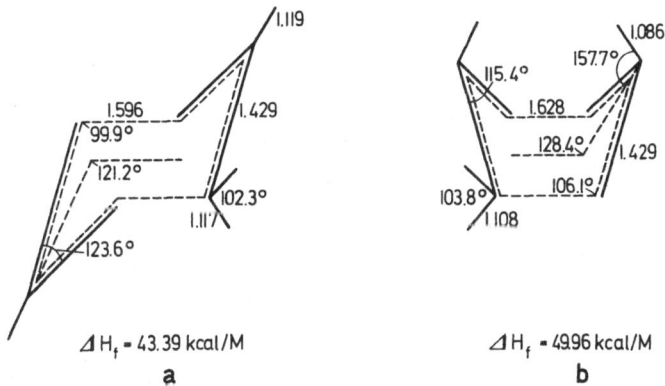

Fig. 5. Calculated geometries and energies (kcal/mole relative to 6) for (a) 9; (b) 8. Reproduced from Ref. [32]

It will be seen that the "chair" transition state *(9)* is predicted to be more stable than the "boat" *(8)* by 6.5 kcal/mole; the experimental evidence implies [29] that 9 is favoured by "not less than 5.7 kcal/mole". The experimental value is based [29] on the presence of a very small amount of product (< 1%) that could have been formed via 8; since the amount was so small, this was identified only by its g.l.c. retention time. If the identification was correct, the ratio of products implies a difference in activation energy of 5.7 kcal/mole between 8 and 9; our calculations suggest that the identification was correct and that Doering and Roth were overcautious.

A very surprising feature of the "boat" transition state is the distortion of the hydrogen atoms at the center of each allyl moiety out of coplanarity with the $C_3$ unit. If the distortion had been outwards, this could have been attributed to steric repulsion; in fact it is *inwards*. Now the interpretation [31] of this reaction in terms of Evans' principle attributes the favouring of 9 to an antibonding interaction in 8 between the $2p$ AOs of the central carbon atoms in the allyl moieties; the distortion of the hydrogen atoms from coplanarity, indicated in Fig. 5b, could reduce this interaction by replacing the two carbon $2p$ AOs by hybrid AOs with less mutual overlap (cf. *10*).

A similar situation arises in the $\pi - \pi^*$ excited states of *ethylene* and *acetylene* where the $\pi$ interactions become antibonding; the excited states are consequently twisted *(11)* and bent *(12)* respectively. Another example is provided by triatomic molecules formed by second row atoms; if the total number of valence electrons exceeds 16, the molecules are bent [33] since there are now antibonding electrons present and bending reduces the resulting unfavourable $\pi$ interaction by replacing $p$ AOs by hybrid AOs (cf. ozone, *13*).

*10*  *11*  *12*  *13*

This interpretation was confirmed [32] by repeating the calculations for 8 and 9, omitting the integrals representing interactions between the $2p$ AOs of carbon $a$ and $b$; the energy difference between the two transition states then vanished and the hydrogen atoms at positions $a$ and $b$ in 8 reverted to the planes of the allyl moieties.

If the transition state has the chair conformation 9, groups in the terminal positions of each allyl moiety can occupy pseudoaxial or pseudoequatorial positions. We therefore calculated the structures with a single methyl subsituent; the results are shown in Fig. 6.

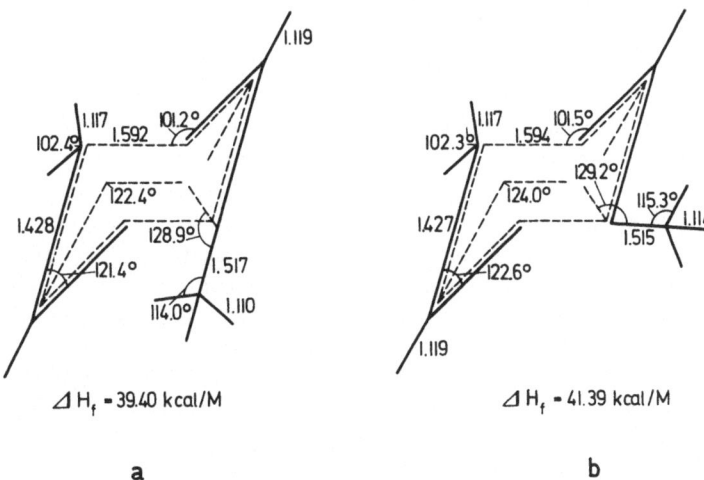

Fig. 6. Geometries and relative energies (kcal/mole) of (a) equatorial and (b) axial methyl derivatives of 9. Reproduced from Ref. [32]

It will be seen that the equatorial conformer is predicted to be favoured by 2.0 kcal/mole; experiment shows it to be indeed favoured by 1.5 kcal/mole [34].

Thus our calculations not only reproduce quantitatively the results of two extensive experimental studies but also show very clearly why it is that 9 is favoured over 8. Since the total cost of our work (1 month by a Postdoctoral Fellow, Dr. Wolfgang Schoeller, *plus* 5 hours time on a CDC 6600 computer) amounted to about 20 000, DM, i.e. a small fraction of the two experimental studies referred to above, one can see that *MO theory can already compete with experiment,* at least in possibly favoured cases as this.

Given that the boat transition state 8 is unfavourable, it is at first sight surprising that the Cope rearrangements of *bullvalene (14), barbaralane (15),* and *semibullvalene (16)* should take place so readily given that the transition states *(17)* of these reactions are derivatives of 8. We therefore decided [35] to calcu-

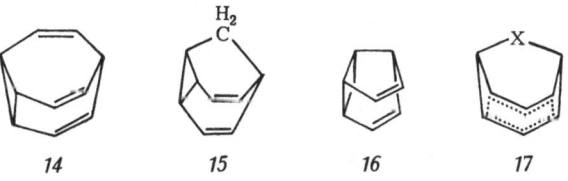

late the activation energies for these processes. Indeed, it seemed likely that our procedure might be even better here than in the case of *6;* for since our method tends to underestimate the energies of cyclic compounds, the difference in energy between *6* and *9* was underestimated (calculated, 24; observed [29] 35 kcal/mole). In the case of *14– 16,* the transition states are so similar to the reactants in geometry that one might expect our procedure to be even more successful.

Table 4 compares calculated [35] and observed activation energies for the Cope rearrangements of *6, 14,* and *15.* It will be seen that our procedure correctly accounts for the great increase in reactivity in going from *6* to bullvalene and barbaralane, and also that our prognostication, that the calculation should be more accurate for *14* and *15,* was correct. The rearrangement of *16* is interesting in that it is fast [36] on the n.m.r. time scale at –110°, leading to the suggestion [36] that in this case the nonclassical intermediate *17* ($X = -$) might be more stable than the classical isomer *16;* indeed, calculations [37] by the "extended Hückel" method predicted this to be the case. Our calculations suggest that *16* is still the more stable form, though only by a small amount; the experimental evidence, while inconclusive, suggests that this is in fact the case [36]. The difference between *16* and *17* is, however, so small that one should certainly be able to displace the balance in favour of *17* by suitable substitution; since our method apparently works so well in this series, and since the calculations can be carried out much more quickly and much more cheaply than

Table 4. *Calculated (MINDO/2) and Observed Activation Energies for Degenerate Rearrangements in the Bullvalene Series*

| Compound | Activation energy for rearrangement (kcal/mole) | |
| --- | --- | --- |
| | calcd. | obsd. |
| Biallyl *(6)* | 30.6[a] | 41.2[a],[b]· |
| Bullvalene *(14)* | 11.3 | 11.8[c]; 12.8[d] |
| Barbaralene *(15)* | 5.9 | 8.6[e] |
| Semibullvalene *(16)* | 3.3 | – |

a) For arrangement via the boat transition state, *8.*
b) Calculated on the assumption that a minor product of the reaction was correctly idenfied; Doering and Roth [29] cautiously quote the difference in energy between *8* and *9* as "greater than 5.7 kcal/mole".
c) M. Saunders, *Tetrahedron Letters,* 1699 (1963).
d) A. Allerhand and H.S. Gutowsky, *J. Amer. Chem. Soc., 87,* 4092 (1965).
e) W. von E. Doering, B. M. Ferrier, E. T. Fossel, J. H. Hartenstein, M. Jones, Jr., G. Klumpp, R. M. Rubin, and M. Saunders, *Tetrahedron, 23,* 3943 (1967).

synthesis of the derivatives in question, this seems an ideal problem for it. Needless to say we are pursuing this actively.

## D. Structures and Reactions of 7-Norbornyl Ions and Radicals

The properties of the *7-norbornyl (18), 7-nobornenyl (19),* and *7-norborna-dienyl (20)* ions and radicals have aroused much interest in view of the evidence that the cations *19* and *20* have nonclassical π-complex structures. We

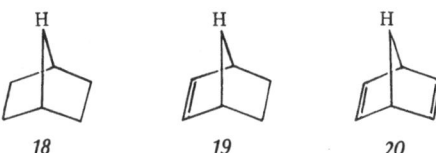

have accordingly calculated [23] all nine species. It is true that MINDO/2 has not been tested in situations of this kind; however its success in dealing with a variety of transition states containing equally „nonclassical" structures seems to justify this extrapolation. Fig. 7 shows the geometries calculated for the nine species; the only assumption made was that in each case there is a plane of symmetry through carbon atom 7 and the midpoints of the 2, 3 and 5, 6 bonds.

The cations *18a* and *19a* are predicted to be distorted, the bridge atom (7) being tilted towards the double bond of *18* or one double bond of *19* and the hydrogen atom tilted away; this is exactly what would be expected if the ions are π complexes [38] in which one double bond forms a dative bond to the 7-carbon atom. It is perhaps surprising that the saturated cation *(18a)* shows a similar deformation; presumably I-strain at the 7-position favours a π-complex-like structure in which a C—C σ bond acts as donor.

The anions *18c* and *19c* show an opposite deformation, the 7-hydrogen now being tilted *towards* the adjacent double bond. In this case the interaction between the lone pair electrons at the 7-position and any filled orbital should be antibonding; presumably these interactions are minimized by tilting the 7-carbon in the way indicated, this tilt reducing the overlap between the lone pair and other bonds in the molecule.

The radicals are intermediate between the cations and anions; one might therefore expect the unpaired electron to show little or no interaction with other bonds. It is therefore not surprising that the radicals are predicted to have symmetrical structures, the 7-carbon atom being planar in each case.

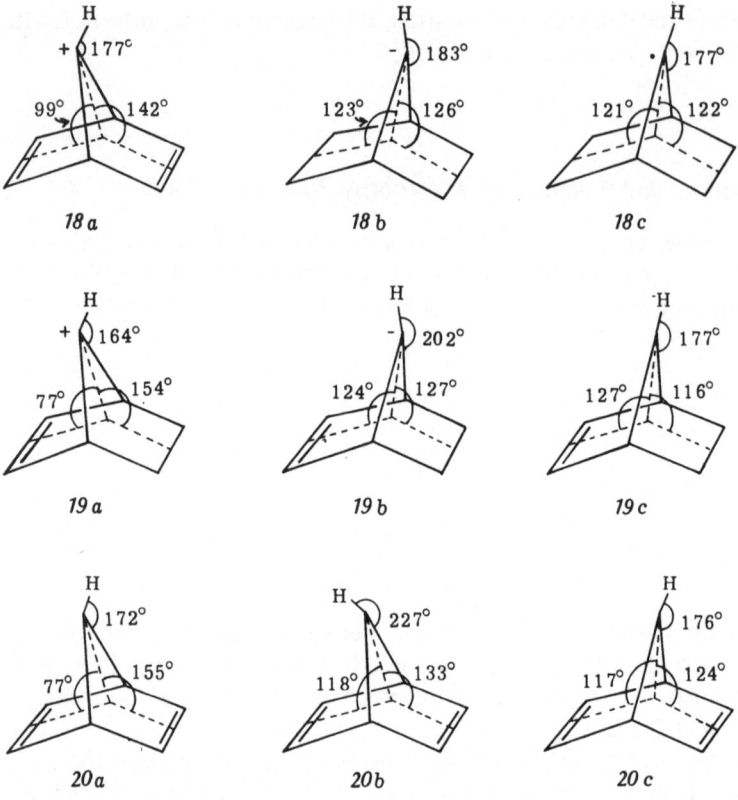

Fig. 7. Calculated geometries for *18, 19, 20*; (a) cation; (b) radical; (c) anion

The distorted structure predicted for *19a* has been confirmed [39] by n.m.r. studies; one might on this basis expect interconversion of the two isomers *20* and *22* via a symmetrical intermediate *21* to take place at a measurable rate.

We have calculated the reaction path for this process; the predicted activation energy is 26 kcal/mole, which is consistent with the available evidence ($\Delta G^+ > 19.6$ kcal/mole [39]).

## E. Electrocyclic Ring Opening of Cyclopropyl Ions and Radicals

*Cyclopropyl ions* and radicals *(23)* can undergo conversion to allyl *(24)* by typical electrocyclic ring opening processes; we have carried out calculations [40] for ring opening by both conrotatory and disrotatory paths. Table 5 shows calculated activation energies for the various processes.

Table 5. *Activation Energies for Ring Opening of Cyclopropyl Ions and Radicals*

| Compound | Calculated activation energy for ring opening (kcal/mole) | |
|---|---|---|
| | disrotatory | conrotatory |
| ▷+ | 7.4 | 38.0 |
| ▷· | 24.7 | 52.3 |
| ▷− | 65.7 | 30.7 |

It will be seen that the cation is predicted to undergo disrotatory ring opening and the anion conrotatory ring opening, in agreement with predictions from the Woodward-Hoffmann rules [30] or Evans' principle [31]. In each case the activation energy for the less favoured path is greater by 30–35 kcal/mole. The calculated reaction paths are also interesting. In the case of the favoured processes, the methylene groups begin to rotate as soon as the bond joining them starts to lengthen, whereas in the disfavoured ones the methylene groups do not begin to rotate until near the end of the reaction, near the maximum in the potential curve, when they suddenly rotate through 90°.

In the case of the radical, disrotatory opening is favoured and by a surprisingly large amount (28 kcal/mole). Moreover both disrotatory and conrotatory processes show the same characteristics as disrotatory opening of the cation or conrotatory opening of the anion, i.e. the methylene groups start to rotate at the beginning of the reaction. According to the first order treatment of aromaticity [31], both paths should be equally favourable in the case of the radical, any difference between them beiing due to second order effects. Recent work has in fact shown[41] that the (4n + 3) Hückel radicals are aromatic rather than nonaromatic as predicted by first order theory [31], and this conclusion is also in agreement with some unpublished calculations [42]. On this basis the cyclopropyl radical should undergo disrotatory ring opening preferentially. It should be noted that extended Hückel calculations [43] led to the opposite conclusion that conrotatory ring opening should be favoured. It is to be hoped that this point will soon be clarified by experiment.

Note that we predict ring opening of the cyclopropyl cation to require activation; this at first sight seems to be at variance with evidence that rearrangement occurs as a concerted process in the solvolysis of cyclopropyl esters and indeed acts as a driving force [44]. Moreover the evidence shows very clearly that this is so only for *one* of the possible disrotatory processes, i.e. that indicated in *25:*

*25*          *26*

A clue to this behaviour was provided by a study of the reaction path for rearrangement of the *cyclopropyl cation.* Initially the carbon atom at the carbonium ion center is coplanar; however as the ring begins to open, the adjacent hydrogen atom tilts out of coplanarity in the sense corresponding to the process indicated in *25.* When the calculations were repeated starting with a nonplanar form *(26)* of the cation, rearrangement took place without activation in the same sense. The ability of this rearrangement to act as a driving force in the solvolysis of cyclopropyl esters is therefore a consequence of the fact that solvolysis produces the nascent carbonium ion in a nonplanar state.

## F. Electrocyclic Ring Opening in Cyclobutene and Bicyclobutane

We are also carrying out calculations [40] for the electrocyclic ring opening of *cyclobutene (27)* and *bicyclobutane (28)* to *1,3-butadiene (29).* While.this work is not yet complete, we have established that *27* opens preferentially by

*27*          *28*          *29*

a conrotatory process while in *28* the opening of one ring is conrotatory, the other disrotatory. These results are in agreement with experiment [45] and with predictions based on Evans' principle [31] and, in the case of *27*, with the Woodward-Hoffmann rules [30]. It is of course impossible to apply arguments based on orbital symmetry to *28* since no symmetry is conserved during the reaction.

## G. Addition of Carbon Atoms to Olefines

Under suitable conditions, carbon atoms react with olefines to form allenes [46)], presumably via intermediate cyclopropyl carbenes; viz.

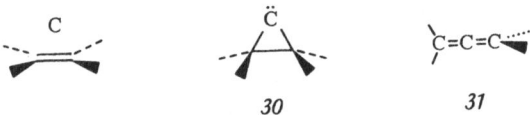

30          31

The evidence suggests [47)] that the carbon atoms must be in an excited singlet state, claimed to be $^1S$; given this, the reactions take place with great facility even in a matrix at $-190\,°C$.

Fig. 8 shows a plot of the calculated [48)] reaction path for the reaction of $^1S$ carbon atoms with *ethylene*. It will be seen that the intermediate carbene *(28)* is formed exothermically, but that its rearrangement to allene *(29)* requires much activation (50 kcal/mole). At first sight this seems inconsistent with the evidence that the reaction takes place readily at $-190\,°C$.

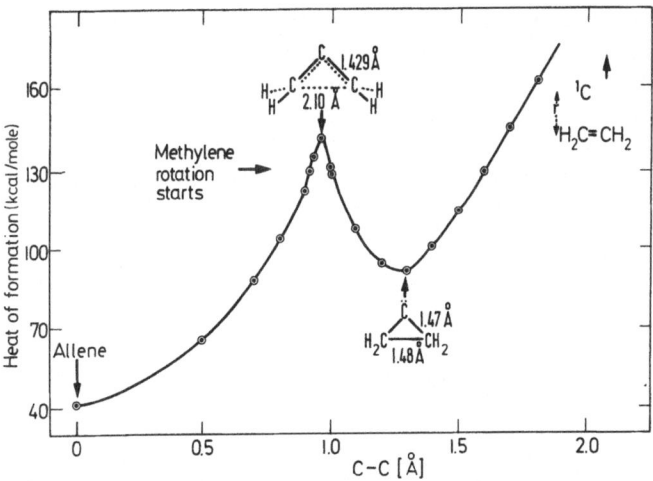

Fig. 8. Reaction path for reaction of a carbon atom with ethylene; reaction coordinate (r) indicated in inset. Reproduced from Ref. [48)]

In order to convert *30* to *31*, one must rotate the terminal methylene groups through *90°* relative to one another because in *30* the four hydrogen atoms are coplanar whereas in *31* they lie in orthogonal planes. According to our calculations [48)], this rotation does not begin until the transition state is passed (see Fig. 8).

The approach of the carbon atom to ethylene, and the conversion of *30* to *31*, thus correspond to one of the normal modes of vibration of the cyclopropane ring, viz:

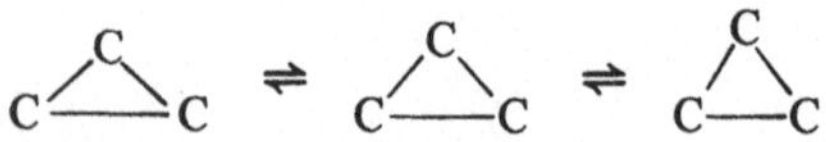

$$\triangle \rightleftharpoons \triangle \rightleftharpoons \triangle$$

Evidently a large part of the energy liberated in the approach of the carbon atom to ethylene will go into this normal mode — which is the one required for conversion of *30* to *31*. Unless the interconversion of vibrational energy is incredibly efficient, one would then expect the initially formed *30* to be converted to *31* even at the lowest temperatures. The fact that allene is formed at −190° is not therefore surprising. On the other hand the existence of a large barrier between *30* and *31* would prohibit rearrangement of *30* if formed under milder conditions; free cyclopropyl carbenes do not rearrange to allenes if formed by conventional methods in solution [49].

## H. Structures and Reactions of Carbenes

We have carried out extensive calculations[23,50] for carbenes with results that again seem to account well for their reported properties. Thus Fig. 9 compares the calculated and observed [51] geometries of singlet and triplet carbene; the agreement is excellent. Note that triplet carbene was originally reported [51] to be linear on the basis of its ultraviolet spectrum; recent e.s.r. studies[52] have shown it to be definitely nonlinear, the HCH angle being approximately 140°.

Fig. 9. Calculated (observed) properties of (a) singlet carbene; (b) triplet carbene. Observed values from G. Herzberg, *Proc. Roy. Soc. (London), A262,* 291 (1961)

The calculated heat of formation of singlet carbene agrees well with experiment. That of triplet carbene is not known. The ionization potential of singlet carbene, determined spectroscopically [51] (10.26 eV) is much less than the electron impact value [53] (11.9 eV); if the latter refers to triplet carbene, as seems likely, the singlet-triplet separation is 1.64 eV or 37.8 kcal/mole, which is quite close to our estimate (28.5 kcal/mole).

Fig. 10 shows calculated geometries and heats of formation of several simple carbenes. Note that *methylcarbene, dimethylcarbene,* and *formylcarbene* are predicted to have nonclassical structures with bridging hydrogen atoms and heats of formation that are more negative than expected from analogy with carbene; this would account well for the fact [54] that neither methylcarbene nor dimethylcarbene undergoes normal insertion or addition reactions, rearranging instead to ethylene or propene, while formylcarbene is much less reactive than other acrylcarbenes and tends instead to rearrange to ketene. The "dicarbene" $C_3$ is predicted to have a triangular geometry, on the assumption, possibly incorrect, that it has a closed shell structure. The predicted heat of formation (214 kcal/mole) agrees quite well with the value reported [55] for $C_3$ (200 ± 10 kcal/mole).

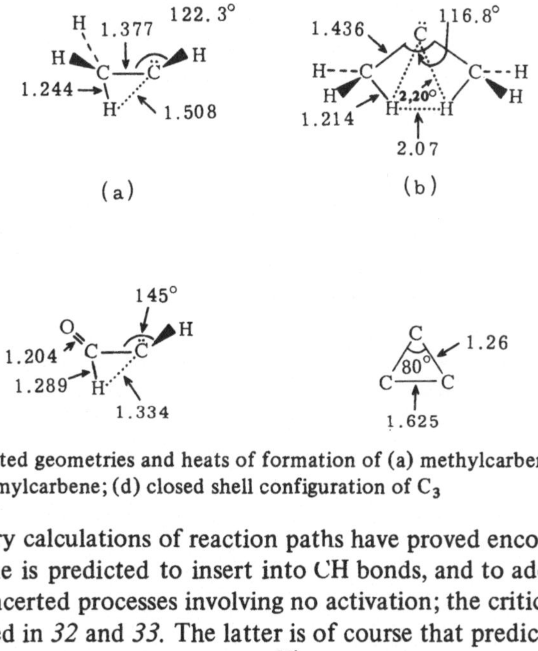

Fig. 10. Calculated geometries and heats of formation of (a) methylcarbene; (b) dimethylcarbene; (c) formylcarbene; (d) closed shell configuration of $C_3$

Preliminary calculations of reaction paths have proved encouraging. Thus singlet carbene is predicted to insert into CH bonds, and to add to double bonds, by concerted processes involving no activation; the critical geometries are as indicated in *32* and *33*. The latter is of course that predicted by Skell [56] and supported experimentally by Closs[57]; it is also in accord with predictions based on considerations of orbital symmetry [30] or Evans' principle [31]. The total lack of discrimination shown by carbene in reactions of this type also indicates that the activation energies must be zero or close to zero.

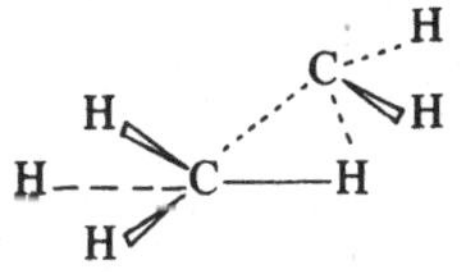

*32*

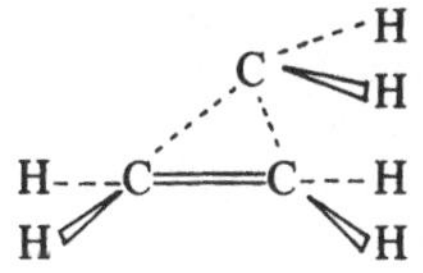

*33*

*Difluorocarbene* is predicted to be much less reactive, concerted insertion into a CH bond of methane involving an activation energy of no less than 26 kcal/mole and even addition to ethylene being activated (5 kcal/mole). Difluorocarbene undergoes insertion into RH bonds only if R is a stabilized radical and then only with loss of configuration, [58] implying that the reaction takes place by abstraction of hydrogen and combination of the resulting pair of radicals (R· + ·CHF$_2$).

# V. Summary and Conclusions

If quantum theory is to be used as a chemical tool, on the same kind of basis as, say, n.m.r. or mass spectrometry, one must be able to carry out calculations of high accuracy for quite complex molecules without excessive cost in computation time. Until recently such a goal would have seemed quite unattainable and numerous calculations of dubious value have been published on the basis that nothing better was possible. Our work has shown that this view is too pessimistic; semiempirical SCF MO treatments, if properly applied, can already give results of sufficient accuracy to be of chemical value and the possibilities of further improvement seem unlimited. There can therefore be little doubt that we are on the threshold of an era where quantum chemistry will serve as a standard tool in studying the reactions and other properties of molecules, thus bringing nearer the fruition of Dirac's classic statement, that with the development of quantum theory chemistry has become an exercise in applied mathematics.

# References

[1] This work was supported by the Air Force Office of Scientific Research through Contract F44620-C-0121.
[2] Dewar, M. J. S., Haselbach, E.: J. Am. Chem. Soc. *92*, 590 (1970).
[3] Roothaan, C. C.: Rev. Mod. Phys. *23*, 29 (1951).
[4] (a) Snyder, L. C., Basch, H.: J. Am. Chem. Soc. *91*, 2189 (1969).
(b) Pople, J. A.: Accounts of Chemical Research *3*, 217 (1970); Hehere, W. J., Ditchfield, R., Radom, L., Pople, J. A.: J. Am. Chem. Soc. *92*, 4796 (1970).
[5] Rabinowitch, R. S., Looney, F. S.: J. Chem. Phys. 23, 2439 (1955).
[6] Hoffmann, R.: J. Chem. Phys. *39*, 1397 (1963).
[7] — J. Am. Chem. Soc. 90, 1475 (1968).
[8] Burr, J. G., Dewar, M. J. S.: J. Chem. Soc. xx. 1201 (1954); Fort, A. W.: J. Am. Chem. Soc. *84*, 4979 (1962).

[9] Pople, J. A., Santry, D. P., Segal, G. A.: J. Chem. Phys. Suppl. *43*, 5129 (1965); Segal, G. A.: J. Chem. Phys. Suppl. *43*, 5136 (1965).

[10] See e.g. Wiberg, K. B., Szeimies, G.: J. Am. Chem. Soc. *92*, 571 (1970).

[11] Fischer, H., Kollmar, H.: Theoret. Chim. Acta *13*, 213 (1969); *16*, 163 (1970); Brown, R. D., James, B. H., O'Dwyer, M. F.: Theoret. Chim Acta *17*, 264 (1970).

[12] Pople, J. A., Beveridge, D. L., Dobosh, P. A.: J. Chem. Phys. *47*, 2026 (1967).

[13] Dewar, M. J. S., Klopman, G.: J. Am. Chem. Soc. *89*, 3089 (1967).

[14] Pople, J. A.: Trans. Faraday Soc. *49*, 1375 (1953).

[15] See Dewar, M. J. S., de Llano, C.: J. Am. Chem. Soc. *91*, 789 (1969); Dewar, M. J. S., Morita, T.: J. Am. Chem. Soc. *91*, 796 (1969); Dewar, M. J. S., Harget, A.: Proc. Roy. Soc. (London) *315*, 443, 457 (1970).

[16] Baird, N. C., Dewar, M. J. S.: J. Chem. Phys. *50*, 1262, 1275 (1969); J. Am. Chem. Soc. *91*, 352 (1969).

[17] Bodor, N., Dewar, M. J. S., Harget, A., Haselbach, E.: J. Am. Chem. Soc. *92*, 3854(1970).

[18] Ohno, K.: Theoret. Chim. Acta, *2*, 219 (1964); Klopman, G.: J. Am. Chem. Soc. *87*, 3300 (1965).

[19] See Allinger, N. L., Hirsch, J. A., Miller, M. A., Tyminski, I. J., Van-Catledge, F. A.: J. Am. Chem. Soc. *90*, 1199 (1968).

[20] This program was written by Sr. Brown, A.

[21] Nelder, J. A., Mead, R.: Computer J. *7*, 308 (1964).

[22] Unpublished work by Dr. Kohn, M.

[23] Unpublished work by Dr. W. W. Scholler.

[24] Hendrickson, J. B.: J. Am. Chem. Soc. *83*, 4537 (1961).

[25] Anet, F. A. L., Bourn, A. J. R.: J. Am. Chem. Soc. *89*, 760 (1967).

[26] Tiers, G. V. D.: Proc. Chem. Soc. 389 (1960).

[27] Dewar, M. J. S., Harget, A., Haselbach, E.: J. Am. Chem. Soc. *91*, 7521 (1969).

[28] (a) Anet, F. A. L.: J. Am. Chem. Soc. *84*, 671 (1962); Anet, F. A. L., Bourn, A. J., Lin, Y. S.: J. Am. Chem. Soc. *86*, 3576 (1964).
(b) Gwynn, D. E., Whitesides, G. M., Roberts, J. D.: J. Am. Chem. Soc. *87*, 2862 (1965).

[29] Doering, W. von, Roth, W. R.: Tetrahedron *18*, 67 (1962).

[30] See Hoffmann, R., Woodward, R. B.: Angew. Chem. Inern. Ed. Engl. *8*, 781 (1969).

[31] Dewar, M. J. S.: Angew. Chem., in press; Tetrahedron Suppl. *8* (1), 75 (1966); "The Molecular Orbital Theory of Organic Chemistry", McGraw-Hill Book Co. Inc., New York, N.Y. (1969).

[32] Brown, A., Dewar, M. J. S., Schoeller, W. W.: J. Am. Chem. Soc. *92*, 5516 (1970).

[33] Walsh, A. D.: J. Chem. Soc. 2266 (1953).

[34] Frey, H. M., Solly, R. K.: J. Chem. Soc. A 1371 (1969).

[35] Dewar, M. J. S., Schoeller, W. W.: J. Am. Chem. Soc., in press.

[36] Zimmermann, H. E., Binkley, R. W., Givens, R. S., Grunewald, G. L., Sherwin, M. A.: J. Am. Chem. Soc. *91*, 3316 (1969).

[37] Personal communication from Professor D. S. Wulfman.

[38] Dewar, M. J. S.: Bull. Soc. Chim. France *18*, C 71 (1951); Dewar, M. J. S., Marchand, A. P.: Ann. Rev. Phys. Chem. *16*, 321 (1965).

[39] Brookhart, M., Lustgarten, R. K., Winstein, S.: J. Am. Chem. Soc. *89*, 6352 (1967).

[40] Unpublished work by S. Kirschner.

[41] Vincow, G., Dauben, H. J., Jr., Hunter, F. R., Volland, W. V.: J. Am. Chem. Soc. *91*, 2823 (1969).

[42] Unpublished work by Dr. C. G. Venier.

[43] Woodward, R. B., Hoffmann, R.: J. Am. Chem. Soc. *87*, 395 (1965).

[44] Schleyer, P. von R., Dine, G. W. van, Schollkopf, U., Paust, J.: J. Am. Chem. Soc. *88*, 2868 (1966).

[45] Closs, G. L., Pfeffer, P. E.: J. Am. Chem. Soc. *90*, 2452 (1968).
[46] MacKay, C., Polack, P., Rosemberg, H. E., Wolfgang, R.: J. Am. Chem. Soc. *84*, 308 (1962); Dubrin, J., MacKay, C., Wolfgang, R.: J. Am. Chem. Soc. *86*, 959 (1964); Marshall, M., MacKay, C., Wolfgang, R.: J. Am. Chem. Soc. p. 4741; Dubrin, J., MacKay, C., Wolfgang, R.: J. Am. Chem. Soc. p. 4747.
[47] Skell, P.S., Engel, R.R.: J. Am. Chem. Soc. *89*, 2912 (1967).
[48] Dewar, M. J. S., Haselbach, E., Shansahl, M.: J. Am. Chem. Soc.
[49] Jones, W. M., Grasley, W. H., Brey, W. S.: J. Am. Chem. Soc. *85*, 2654 (1963).
[50] Unpublished work by Dr. J. Wasson.
[51] Herzberg, G.: Proc. Roy. Soc. (London) *A 262*, 291 (1961).
[52] Personal communication from Dr. Wasserman.
[53] Field, F., Franklin, J. E.: "Electronic Impact Phenomena", Academic Press, New York, p. 117, 1957.
[54] See Kirmse, W.: "Carbene Chemistry".
[55] Chupka, W. A., Inghram, M. G.: J. Phys. Chem. *59*, 100 (1955).
[56] Skell, P. S. , Garner, A. Y.: J. Am. Chem. Soc. *78*, 5430 (1956); Skell, P.S. and Woodworth, R. C.: J. Am. Chem. Soc. *78*, 4496 (1956).
[57] Closs, G. L., Closs, L. E.: Angew. Chem. *74*, 431 (1962).
[58] Franzen, V.: Unpublished results (1962).

Received December 15, 1970

# Localized Molecular Orbitals:
# A Bridge between Chemical Intuition
# and Molecular Quantum Mechanics*

**Walter England, Lydia S. Salmon**, Klaus Ruedenberg**

Institute for Atomic Research and Department of Chemistry Iowa State University, Ames, Iowa, USA

## Contents

*) Work performed in the Ames Laboratory of the U.S. Atomic Energy Commission, Contribution No. 2816.

**) Present address: Dept. of Chemistry, Washington State University, Pullman, Washington 99163, USA.

## Introduction

Since rigorous theoretical treatments of molecular structure have become more and more common in recent years, there exists a definite need for simple connections between such treatments and traditional chemical concepts. One approach to this problem which has proved useful is the method of localized orbitals. It yields a clear picture of a molecule in terms of bonds and lone pairs and is particularly well suited for comparing the electronic structures of different molecules. So far, it has been applied mainly within the closed-shell Hartree-Fock approximation, but it is our feeling that, in the future, localized representations will find more and more widespread use, including applications to wavefunctions other than the closed-shell Hartree-Fock functions.

The following presentation is limited to closed-shell molecular orbital wavefunctions. The first section discusses the unique ability of molecular orbital theory to make chemical comparisons. The second section contains a discussion of the underlying basic concepts. The next two sections describe characteristics of canonical and localized orbitals. The fifth section examines illustrative examples from the field of diatomic molecules, and the last section demonstrates how the approach can be valuable even for the "delocalized" electrons in aromatic $\pi$-systems. All localized orbitals considered here are based on the self-energy criterion, since only for these do the authors possess detailed information of the type illustrated. We plan to give elsewhere a survey of work involving other types of localization criteria.

## 1. Chemistry and Comparison

Even though the understanding of individual molecules is an important aspect of chemistry, actual chemical research proceeds through *comparison* of molecular systems. Important information is gained from measuring energy differences of various systems, such as reaction energies, ionization energies or excitation energies. Even in the case of properties which pertain to isolated systems, such as dipole moments, the chemist is particularly interested in comparing them for series of related compounds in order to establish trends and regularities. This method has a noteworthy consequence for the theory of a quantum mechanical observable. To say that several systems are related is to say that it is possible to identify subsystems common to all of them. If this division of each system into a common part and a different part is reflected in an appropriate manner by the quantum mechanical treatment, then it can become possible to relate theoretical quantities characteristic of the subsystems to differences or other algebraic combinations of experimental quantities of the related systems themselves. *The comparison of related molecular systems can therefore yield experimental information pertaining to theoretical quantities of subsystems, such as orbital contributions, which, because of the indistinguishability of electrons, would not qualify as observables in the sense of being expectation values of an operator for any one isolated system.*

In this "research by comparison" two fundamentally different approaches can be distinguished: On the one hand, the investigator may compare the same molecule in various quantum states. The entire field of spectroscopy is based on this type of measurement. On the other hand, he may compare corresponding states, e.g. ground states, of structurally related molecules in order to gain insights of importance for understanding chemical reactions. Until recently it was a common belief that quite different theoretical models were needed for bringing order into the variety of phenomena observed in these two areas of comparing chemical systems. Ever since Robert S. Mulliken explained the band spectra of diatomics with the help of the molecular orbital theory [1], there has been no question that this was the suitable approach for making sense of molecular spectroscopy; and ever since Linus Pauling wrote "The Nature of the Chemical Bond", [2] it has been taken for granted that the most fruitful comparison of ground states of different molecules could be obtained with the help of valence bond ideas. Unfortunately, it has not been possible to combine these two approximations without destroying their conceptual simplicity.

The development of localized-orbital aspects of molecular orbital theory can be regarded as a successful attempt to deal with the two kinds of comparisons from a unified theoretical standpoint. [3] It is based on a characteristic flexibility of the molecular orbital wavefunction as regards the choice of the molecular orbitals themselves: the same many-electron Slater determinant can be expressed in terms of various sets of molecular orbitals. In the classical spectroscopic approach one particular set, the *canonical* set, is used. On the other hand, for the same wavefunction an alternative set can be found which is especially suited for comparing corresponding states of structurally related molecules. This is the set of *localized* molecular orbitals. Thus, it is possible to cast *one* many-electron molecular-orbital wavefunction into *several* forms, which are adapted for use in different comparisons: *for a comparison of the ground state of a molecule with its excited states the canonical representation is most effective; for a comparison of a particular state of a molecule with corresponding states in related molecules, the localized representation is most effective.* In this way the molecular orbital theory provides a unified approach to both types of problems.

What has been said applies to approximate as well as to *ab-initio* molecular orbital wavefunctions, i.e. those obtained by solving the self-consistent-field equations exactly. Hence, the localized orbital approach also offers an attractive tool for bridging the gap between rigorous quantitative calculations and qualitative chemical intuition. The experience gained so far has shown that interpretations suggested by the localized orbital picture correspond closely to intuitive chemical thinking.

## 2. Flexibility of Molecular Orbitals

### 2.1 Total Wavefunction

For the purpose of this presentation we limit ourselves to closed-shell systems, i.e. those systems consisting of an even number, $2N$, of electrons which doubly occupy $N$ space orbitals $u_1, u_2, \ldots u_N$, each with $\alpha$ and $\beta$ spin. The appropriate molecular orbital wavefunction is then given by

$$\psi = \mathcal{A} \{(u_1\alpha)^{(1)} (u_1\beta)^{(2)} \ldots (u_N\alpha)^{(2N-1)}(u_N\beta)^{(2N)}\}, \tag{1}$$

where the superscripts denote the electrons and, e.g.,

$$(u_k\alpha)^{(j)} = u_k(j)\alpha(j). \tag{2}$$

Moreover $\mathcal{A}$ represents the antisymmetrizer

$$\mathcal{A} = (N!)^{-1/2} \sum_P (-1)^P P, \tag{3}$$

where the summation extends over all permutations $P$ among the $N$ electrons. In his original paper on the Hartree-Fock equations, Fock pointed out [4] that the very same $N$ electron wavefunction can be expressed in terms of a different set of space orbitals, $v_1, v_2 \ldots v_N$, using the same Ansatz, namely,

$$\psi = \mathcal{A} \{(v_1\alpha)^{(1)} (v_1\beta)^{(2)} \ldots (v_N\alpha)^{(2N-1)} (v_N\beta)^{(2N)}\}. \tag{4}$$

The only condition for the expressions (1) and (4) to represent the same wavefunction is that the space orbitals $u_k$ and the space orbitals $v_k$ are related by a linear transformation

$$v_k(x) = \sum_{i=1}^{N} u_i(x) T_{ik}, \tag{5}$$

with $T$ being an orthogonal matrix, i.e.

$$\sum_i T_{ik} T_{ij} = \sum_i T_{ji} T_{ki} = \delta_{jk}. \tag{6}$$

For small determinantal wavefunctions these statements are easily verified by explicit expansion; the general proof rests on the fact that the determinant of a matrix product is equal to the product of the determinants of the matrices.

In the present discussion, the molecular orbitals $u_k$ are assumed to form a real, orthonormal set, i.e.

$$\int dV \, u_i(x) u_k(x) = \delta_{ik}. \tag{7}$$

By virtue of the orthogonality of the transformation matrix $T$, the orbitals $v_k$ will then also form an orthonormal set. Nonorthogonal molecular orbitals could be used and, furthermore, transformations with a nonorthogonal (but non-singular) matrix $T$ would be acceptable in Eq. (5). This would introduce an additional factor [Determinant $(T)]^{-2}$ on the right hand side of Eq. (4). For reasons which will become clear later, we limit ourselves to orthogonal molecular orbitals and all explicit formulae to be quoted are specific for this case.

The identity of Eqs. (1) and (4) is expressed by the statement: *The molecular orbital wavefunction (1) is invariant under the orbital transformations given by Eqs. (5) and (6).*

## 2.2 Density

The wavefunction $\psi$ gives rise to the one-electron density given either by

$$\rho(x) = 2 \sum_n u_n^2(x) \tag{8}$$

or by

$$\rho(x) = 2 \sum_n v_n^2(x). \tag{9}$$

The equality of the right-hand sides of Eqs. (8) and (9) is readily verified with the help of Eqs. (5) and (6). More generally one has the identity

$$\rho(x, x') = 2 \sum_n u_n(x) u_n(x') = 2 \sum_n v_n(x) v_n(x') \tag{10}$$

which defines the "density matrix". Thus, *the density matrix and in particular the one-electron density are invariant under the orbital transformation given by Eqs. (5) and (6).*

## 2.3 Energy

The hamiltonian of the system is assumed to have the form

$$\mathcal{H} = \sum_\nu h_\nu + \sum_{\nu < \mu} r_{\nu\mu}^{-1} \tag{11}$$

where $h_\nu$ contains all one-electron operators acting on electron $\nu$, and $r_{\nu\mu}^{-1}$ is the repulsion between electrons $\nu$ and $\mu$. Atomic units are adopted (length: Bohr radius $= a$, energy: Hartree unit $= e^2/a$). With this hamiltonian, the energy expectation value of the wavefunction $\psi$ can be expressed as [5]

$$E = (\psi | \mathcal{H} | \psi) = H + C - X, \tag{12}$$

where the term $H$ represents the "one-electron energy"

$$H = 2 \sum_n (u_n | h | u_n) = 2 \sum_n (v_n | h | v_n), \tag{13}$$

the term $C$ represents the "Coulombic part" of the electron repulsion energy:

$$C = 2 \sum_n \sum_m [u_n^2 | u_m^2] = 2 \sum_n \sum_m [v_n^2 | v_m^2], \tag{14}$$

and the term $X$ represents the "Exchange part" of the electron repulsion energy:

$$X = \sum_n \sum_m [u_n u_m | u_n u_m] = \sum_n \sum_m [v_n v_m | v_n v_m]. \tag{15}$$

In Eqs. (14) and (15) the symbol $[f|g]$ denotes the electrostatic repulsion integral

$$[f|g] = \int dV_1 \int dV_2 f(x_1) g(x_2)/|x_1 - x_2|. \tag{16}$$

The equality between the expressions involving the $u_k$ and those involving the $v_k$, in Eqs. (13), (14), (15), is again verified with the help of the relations (5) and (6).

It is thus found that the *one-electron energy H, the Coulomb energy C, and the Exchange energy X are separately invariant under the orbital transformation given by Eqs. (5) and (6).*

## 2.4. Self-Consistent-Field Equations

The results discussed so far are a consequence of the *determinantal form* of the molecular orbital wavefunction $\psi$. They are valid for an arbitrary choice of molecular orbitals $u_k$. Of particular interest are, however, those space orbitals which render $\psi$ optimal, i.e. that molecular orbital (MO) approximation which is closest to the true wavefunction. The optimal molecular orbitals are those which minimize $(\psi | \mathcal{H} | \psi)$, and Fock [6] has shown that they are the solutions of the integro-differential equations

$$\mathcal{F} u_k(x) = \sum_j u_j(x) \lambda_{jk} \tag{17}$$

known as Hartree-Fock or Self-Consistent-Field (SCF) Equations. Here $\mathcal{F}$ is the Fock operator given by

$$\mathcal{F} = h + \mathcal{C} - \mathcal{X}, \tag{18}$$

where $h$ is the one-electron operator occurring in Eq. (11) and $\mathcal{C}$ and $\mathcal{X}$ are the Coulomb and exchange operators respectively, defined by

$$\mathcal{C} f(x) = \{ \int dV' \rho(x')/|x - x'| \} f(x), \tag{19}$$

$$\mathcal{X} f(x) = (1/2) \int d V' \rho(x, x') f(x')/|x - x'|. \tag{20}$$

For the present purpose, the sole point of interest is that $\mathcal{C}$ and $\mathcal{X}$, and hence $\mathcal{F}$, depend upon the molecular orbitals only via the density matrix defined in Eqs. (8), (9), (10). Hence one can write

$$\mathcal{F} = \mathcal{F}(\rho) = \mathcal{F}\{2 \sum_n u_n(x)u_n(x')\} = \mathcal{F}\{2 \sum_n v_n(x)v_n(x')\}. \qquad (21)$$

Thus, *the Fock-operator is also invariant under the orbital transformations given by Eqs. (5) and (6).*

The $\lambda_{ij}$ are Lagrangian multipliers arising from the side conditions of Eq. (7) which maintain orbital orthonormality during the minimization process. [7] Solution of Eq. (17) in conjunction with Eq. (7) determines *simultaneously* the $u_k(x)$ and the $\lambda_{jk}$. Thus, the $u_k(x)$ *and the* $\lambda_{jk}$ must be considered as the unknowns in the integro-differential equation (17).

By virtue of Eqs. (5) and (6), one can derive from the SCF equations for the $u_k$, i.e. Eq. (17), the following SCF equations for the $v_k$:

$$\mathcal{F} v_k(x) = \sum_j v_j(x) \, \hat{\lambda}_{jk} \, , \qquad (22)$$

where

$$\hat{\lambda}_{jk} = \sum_{rs} \lambda_{rs} T_{rj} T_{sk} \, , \qquad (23a)$$

$$\hat{\lambda} = T^{\text{transposed}} \lambda T. \qquad (23b)$$

Thus, the orbitals $u_k$ and $v_k$ satisfy Hartree-Fock equations which are identical in form and differ only in the numerical values of the constants $\lambda_{jk}$ and $\hat{\lambda}_{jk}$ respectively. But since the latter are unknowns in the equation, and since $\mathcal{F}(\rho)$ is itself invariant as shown in Eq. (21), we can say that *the Hartree-Fock self-consistent-field equations are invariant under the orbital transformation given by Eqs. (5) and (6).* This means in effect, that the energy integral $(\psi \,|\, \mathcal{H} \,|\, \psi)$ is minimized by the $v_k$'s as well as by the $u_k$'s – a circumstance which is in agreement with the invariance of $\psi$ and $(\psi \,|\, \mathcal{H} \,|\, \psi)$ under the transformation (5).

## 2.5. Degrees of Freedom

In view of the preceding considerations it should be emphasized that *it is incorrect to talk about "the" self-consistent-field molecular orbitals of a molecular system in the Hartree-Fock approximation.* The correct point of view is to *associate the molecular orbital wavefunction* $\psi$ *of Eq. (1) with the N-dimensional linear Hilbert space spanned by the orbitals* $u_1, u_2, \ldots u_N$: *any set of N linearly independent functions in this space can be used as molecular orbitals for forming the antisymmetrized product.*

As noted earlier, we limit ourselves arbitrarily, but judiciously, to *orthonormal* orbital sets in this function space, which implies the orthogonality conditions of Eq. (6). This equation represents $1/2 \, N(N + 1)/2$ conditions for the $N^2$ matrix elements of $T$. Thus an orthogonal transformation of degree $N$ contains $N(N - 1)/2$ arbitrary parameters. Hence *there exist N(N - 1)/2 de-*

*grees of freedom in choosing orthonormal self-consistent-field orbitals in the N-dimensional Hilbert space associated with $\psi$.*

## 2.6. Symmetry Properties

If there is a molecular symmetry group whose elements leave the hamiltonian $\mathcal{H}$ invariant, then the closed-shell wavefunction $\psi$ belongs to the totally symmetric representation of both the spin and symmetry groups.[8] It is further true that under these symmetry operations the molecular orbitals transform *among each other* by means of an orthogonal transformation, such as mentioned in Eq. (5) [9] and, therefore, span a representation of the molecular symmetry group. In general, this representation is *reducible.*

# 3. Canonical Molecular Orbitals

## 3.1. Hartree-Fock Equations

By virtue of the orthogonality conditions of Eq. (7), one derives from the SCF Eq. (17) the expression

$$\lambda_{ji} = \lambda_{ij} = (u_i | \mathcal{F} | u_j) \tag{24}$$

for the Lagrangian multipliers $\lambda_{ij}$. Since we have seen that the molecular orbitals $u_k$ contain $N(N-1)/2$ arbitrary parameters, it stands to reason that an equal number of the Lagrangian multipliers can be given arbitrary values (at least within certain limits). This applies in particular to the $N(N-1)/2$ off-diagonal Lagrangian multipliers. One possibility is to require that all off-diagonal Lagrangian multipliers vanish, i.e.

$$\lambda_{ij} = 0 \quad \text{for } i \neq j. \tag{25}$$

These conditions determine a unique set of molecular orbitals, the *canonical molecular orbitals,* (CMO's), $\phi_n$. Inserting the conditions (25) in the SCF Eqs. (17), one sees that the CMO's are solutions of the canonical Hartree-Fock equations [10]

$$\mathcal{F} \phi_n(x) = \epsilon_n \phi_n(x). \tag{26}$$

Since the Hartree-Fock operator $\mathcal{F}$ is readily seen to be Hermitian, it is apparent from Eq. (26) that the CMO's necessarily form an orthonormal set.

If one has determined the operator $\mathcal{F}$ by a method which does not simultaneously determine the CMO's, then Eq. (26) can be looked upon as a one-electron Schroedinger equation to be solved for the CMO's. In this sense, the Fock operator can be thought of as an *effective* one-electron hamiltonian. Thus, a one-electron variational problem can be set up: namely, we require that the variation in the expectation value of the Fock operator be stationary, with the side condition that the wavefunction be normalized, i.e.

$$\delta(\phi \mid \mathcal{F} \mid \phi) = 0, \text{ while } (\phi \mid \phi) = 1, \qquad (27)$$

where $\mathcal{F}$ is considered to be fixed. This variational problem leads in the usual manner to Eq. (26) [11].

## 3.2. Symmetry Properties

Since the Hartree-Fock wavefunction $\psi$ belongs to the totally symmetric representation of the symmetry group of the molecule, it is readily seen that the density matrix of Eq. (10) is invariant under all symmetry operations of that group, and the same holds, therefore, for the Hartree-Fock operator $\mathcal{F}$. In this case, it can be proved that the canonical SCF orbitals, being solutions of Eq. (26), are *symmetry orbitals*, i.e. that they belong to *irreducible* representations of the symmetry group. [12] If the number of molecular orbitals is larger than the dimension of the largest irreducible representation of the symmetry group, it must then be concluded that the set of *all N* molecular orbitals form a *reducible* representation of the group which is the direct sum of all the irreducible representations spanned by the CMO's.

## 3.3. Relation to Experiment

Koopmann's theorem establishes a connection between the molecular orbitals of the 2N-electron system, just discussed, and the corresponding $(2N-1)$-electron system obtained by ionization. The theorem states: If one expands the $(2N-1)$ molecular spin-orbitals of the ground state of the *ionized* system in terms of the 2N molecular spin-orbitals of the ground state of the neutral system, then one finds that the orbital space of the ionized system is spanned by the $(2N-1)$ *canonical* orbitals with the lowest orbital energies $\epsilon_k$; i.e. to this approximation the *canonical self-consistent-field orbital with highest orbital energy is vacated upon ionization.* This theorem holds *only* for the canonical SCF orbitals. [13]

At this point it should be noted that, in addition to the $\phi_n$ discussed previously, the canonical Hartree-Fock equations (26) have additional solutions with higher eigenvalues $\epsilon_n$. These are called virtual orbitals, because they are unoccupied in the 2N-electron ground state SCF wavefunction $\psi$. They are orthogonal to the N-dimensional orbital space associated with this wavefunction.

For physical reasons which are related to Koopmann's theorem, it is found that good approximations to the excited states of the molecule in question are obtained when one or more electrons are promoted from certain *canonical* orbitals which are occupied in the ground state to certain virtual *canonical* orbitals. In other words, the orbital energy levels whose occupations are used to interpret the electronic spectra must be those of the occupied and unoccupied *canonical* molecular orbitals, if reasonable agreement with experiment is to be obtained. [13]

According to basic quantum mechanical principles, the "observables" of a system are always expectation values and/or eigenvalues of operators which are symmetric in all electrons ("indistinguishability"). It is obvious therefore, that energies of individual orbitals cannot be observables of a system in a particular state, and this argument is sometimes advanced to contest the usefulness of any discussion in terms of orbitals, either canonical or localized. But we have seen in the foregoing that canonical orbital energies are nonetheless related to experimentally observable ionization potentials and excitation energies. This contradiction is only apparent: the experimental quantity to which the orbital energy is related is the difference in the energies of two systems, atom and ion, or ground state and excited state, and the two systems in question approximately differ in one orbital only. Thus, *while properties of individual orbitals are indeed not expectation values of any one state of any one system, they can nonetheless be related to differences in expectation values of different states or different systems.*

# 4. Localized Molecular Orbitals

## 4.1. Transferability

In forming localized molecular orbitals, (LMO's), the underlying objective is to confine each molecular orbital to as small a space in the molecule as possible and, at the same time, to have these confined molecular orbitals as far removed from each other as possible. The more the orbitals can be confined, and the more they can be separated from each other, the less likely they are to change when distant parts of the molecule are modified. Thus, the more the orbitals are localized in this sense, the more they can be expected to be *transferable* among molecules having related structures. It is this transferability of localized orbitals which makes them appropriate tools for comparing corresponding states of related molecules and for pinpointing differences between them.

Quantitative similarities of molecules can easily be recognized if it is possible to define quantities for molecular parts which are *additive as well as transferable.* Such quantities can be derived from transferable molecular orbitals because any one-electron property, such as dipole moment, quadrupole moment, kinetic energy, is a sum of the corresponding contributions from all molecular orbitals in a system, *if such orbitals are chosen mutually orthogonal.* Thus, for each transferable orthogonal molecular orbital there exists, e.g., a transferable orbital dipole moment. Since chemists appreciate additive decompositions of

41

molecular properties into transferable orbital contributions, there exists a clear interest in mutually orthogonal localized molecular orbitals.

If it proves possible to establish such decompositions of molecular observables in terms of transferable contributions from localized orbitals, then by measuring the observable in question for sufficiently many molecules, it would be possible to deduce quantitative values for the contributions from individual localized orbitals. Thus, we see again that properties of individual molecular orbitals may become accessible to experimental observation *via* the comparison of measurements from various systems.

## 4.2. Localization Criterion

In order to construct localized orbitals for molecules, it is necessary to define a "measure for the degree of localization" of an arbitrary set of molecular orbitals. The "localized orbitals" are then defined as that set of orthogonal molecular orbitals obtained by a transformation of the type given in Eq. (5), for which the measure of localization has the maximum value. It is clear that the resulting localized orbitals will depend, at least to some degree, upon the choice of the localization measure. In the present work the localized molecular orbitals are defined as those self-consistent-field orbitals which maximize the "localization sum" [14)

$$L = \sum_n [u_n^2 | u_n^2] . \tag{28}$$

By virtue of the definition of Eq. (16) it is recognized that this expression represents the sum of the "self-energies" of all occupied molecular orbitals. The higher the self-energy of a particular orbital, the smaller the space to which this orbital is confined. The localization sum of Eq. (28) represents, therefore, an average measure for the degree of concentration of all orbitals in the set.

Furthermore, because of the invariance of the exchange energy, exhibited in Eq. (15), maximization of the localization sum of Eq. (28) implies the simultaneous minimization of the interorbital exchange repulsions

$$\sum_n \sum_{\substack{m \\ n \neq m}} [u_n u_m | u_n u_m] . \tag{29}$$

That is, it reduces the overall local overlap between the various molecular orbitals as much as possible. Finally, because of the invariance of the Coulombic energy, exhibited in Eq. (14), maximization of the localization sum of Eq. (28) also implies the simultaneous minimization of the interorbital Couombic repulsions

$$\sum_n \sum_{\substack{m \\ n \neq m}} [u_n^2 | u_m^2] . \tag{30}$$

That is, it also reduces the long-range repulsions between the orbitals as much as possible. Thus, this localization method achieves three objectives: *concentration of the molecular orbitals, short-range separation of different orbitals, and long-range separation of different orbitals.*

These three objectives could also be achieved if, in the localization sum of Eq. (28), the function $r_{12}^{-1}$ were replaced by any other monotonically varying function of $r_{12}^{-1}$. The choice of $r_{12}^{-1}$ has a further advantage, however. According to Eqs. (12), (14) and (15), the electron repulsion energy can be written as

$$C - X = \sum_n [u_n^2 | u_n^2] + \sum_{\substack{n \\ n \neq m}} \sum_m \{2[u_n^2 | u_m^2] - [u_n u_m | u_n u_m]\} , \qquad (31)$$

where the terms with $n = m$ have been separated from those with $n \neq m$. Let us assume, for the moment, that by an appropriate choice of the molecular orbitals the interorbital exchange interactions in Eq. (29) can be reduced to zero. Then this term would vanish for the localized molecular orbitals and the electron repulsion energy would have the form

$$\sum_n [u_n^2 | u_n^2] + 2 \sum_{\substack{n \\ (n \neq m)}} \sum_m [u_n^2 | u_m^2] . \qquad (32)$$

This expression is just the one which obtains for the Hartree product wavefunction. The difference between this Hartree wavefunction and the Fock wavefunction of Eq. (1) is the absence of the antisymmetrizer $\mathcal{A}$ in that equation. This means that *in the Hartree wavefunction each electron can be identified with a specific molecular orbital,* whereas *in the Fock wavefunction all electrons make use of all orbitals.* The Hartree wavefunction is of course not a proper quantum mechanical wavefunction, since it is not antisymmetric in the electrons. Moreover, for the Fock wavefunction, it is in general not possible to reduce the interorbital exchange energy to zero. But the localized molecular orbitals, as defined here, represent that set of molecular orbitals for which the energy expression comes closest to the Hartree form, i.e. they *come closest to being identifiable with electrons which are not exchanged among different orbitals.*

### 4.3. Localization Equations

It can be shown that the LMO's which maximize the localization sum of Eq. (28) satisfy the set of equations [15)]

$$[\psi_m \psi_n | \psi_m^2] = [\psi_m \psi_n | \psi_n^2] . \qquad (33)$$

These $N(N - 1)/2$ equations uniquely determine the $N(N - 1)/2$ parameters in the orthogonal transformation discussed in Section 2.5. Thus the localized orbitals represent a well defined orthonormal set among all those sets which yield the same $N$ electron wave function $\psi$ of Eq. (1). Furthermore, it can be shown that the localized *self-consistent-field* molecular orbitals satisfy the following SCF equations [16]

$$(\mathcal{F} + \mathcal{G}) \, \psi_n = \eta_n \psi_n \tag{34}$$

where the operator $\mathcal{G}$ is defined by

$$\mathcal{G} f(x) = \int dV' G(x, x') f(x') \tag{35}$$

with

$$G(x, x') = \sum_i \sum_k \psi_i(x) \psi_k(x') \{ \, | \, [\psi_i \psi_k \, | \, \psi_i^2 - \psi_k^2] \, | - \int dV \, \psi_i \mathcal{F} \psi_k \} \tag{36}$$
$$(i \neq k)$$

It should be noted that both of Eqs. (33) and (34) are also satisfied by those molecular orbitals which *minimize* the localization sum.

The Eqs. (33) and (34) could be used for a practical determination of the localized orbitals. So far, however, a different procedure has been used which is based on the premise that the canonical orbitals are determined first. From these, the localized orbitals are then obtained by a sequence of $2 \times 2$ orthogonal transformations which iteratively increase the localization sum until it reaches the maximum.[17]

### 4.4. Relation to Canonical Orbitals

There exists no uniformity as regards the relation between localized orbitals and canonical orbitals. For example, if one considers an atom with two electrons in a $(1s)$ atomic orbital and two electrons in a $(2s)$ atomic orbital, then one finds that the localized atomic orbitals are rather close to the canonical atomic orbitals, which indicates that the canonical orbitals themselves are already highly, though not maximally, localized.[18] (In this case, localization essentially diminishes the $(1s)$ character of the $(2s)$ orbital.) The opposite situation is found, on the other hand, if one considers the two inner shells in a homonuclear diatomic molecule. Here, the canonical orbitals are the molecular orbitals $(1\sigma_g)$ and $(1\sigma_u)$, i.e. the bonding and the antibonding combinations of the $(1s)$ orbitals from the two atoms, which are completely delocalized. In contrast, the localization procedure yields two localized orbitals which are essentially the inner shell orbital on the first atom and that on the second atom.[19] It is thus apparent that the canonical orbitals may be identical with the localized orbitals, that they may be close to the localized orbitals, that they may be identical with the completely delocalized orbitals, or that they may be intermediate in character.

It should be noted that, in the ground state, the lowest *canonical* orbitals are occupied and that, in excited states, some ground state *canonical* orbitals are vacated and, instead, some of the excited *canonical* orbitals are occupied. Thus the ground state as well as the excited states can be described by occupying certain orbitals out of one and the same set of *canonical* orbitals. In contrast, the localized orbitals obtained by localizing the set of orbitals occupied in one state will usually be quite different from those obtained by localizing the canonical orbitals occupied in another state. Thus the localized orbital structure looks very different in different states of a molecule. An example will be given below for the NH molecule. The difference in the localized structures of various states of a molecule is well known from the valence-bond theory. However, in contrast to the latter, the present theory permits one to relate different states in spite of this difference in localized structure: the localized orbitals in any one state can be transformed to canonical orbitals, which constitute a subset of the entire set of canonical orbitals for the molecule. In this way the possibility of transforming between the canonical and localized orbitals permits one to reconcile seemingly different aspects of related molecular wave functions.

An ingenious application of the equivalence between the canonical and localized representations has recently been made by H. B. Thompson. [20] He pointed out that semi-empirical rules pertaining to the geometrical distributions of electron pairs in molecules, in particular those formulated with success by R. G. Gillespie, [21] can be interpreted as rules pertaining to the geometrical arrangements of localized molecular orbitals. Furthermore he pointed out that, given a set of such localized orbitals in a symmetric molecule, it is frequently possible to predict the occupied canonical orbitals by forming those orthogonal linear combinations of the localized orbitals which span irreducible representations of the molecular symmetry group. This is usually possible without numerical calculation. Using this approach, he was able to demonstrate that, for three- and four-atomic molecules, *Gillespie's rules covering localized orbitals are equivalent to certain rules pertaining to the occupancies of canonical orbitals which A. D. Walsh* [22] *had established* from quite different premises, and shown to be in accord with spectroscopic information.

### 4.5. Uniqueness

While the canonical orbitals of a system are unique, aside from degeneracies due to multidimensional representations, this is not always the case for localized orbitals, and there may be several sets of localized orbitals in a particular molecule. This situation is related to the fact that the localization sum of Eq. (28) may have several relative maxima under suitable conditions. If one of these maxima is considerably higher than the others, then the corresponding set of molecular orbitals would have to be considered as "the" localized set. In some cases, however, the two maxima are equal in value, so that there exist two sets of localized orbitals with equal degree of localization. [23] In such a case there is frequently a symmetry operation of the molecule which transforms one set of localized orbitals into the other set of localized orbitals. On the other hand, even in the absence of symmetry, there may be cases where there exist two relative maxima of the localization sum which are close in value and, in such a case, one has to admit the existence of two sets of localized orbitals which are not symmetry related to each other.[24] An interesting situation is found upon localizing the pi-electrons in benzene: a one-parametric family of infinitely many sets of equivalent localized orbitals is found to exist, all of which yield the same, maximal value for the localization sum.[25] The trigonally equivalent lone pairs on each F atom in $F_2$ provide another example: the lone pairs on one atom can be rotated with respect to those on the other without changing the localization sum.[25]

## 4.6. Symmetry Properties

There exists no uniformity as regards the relations between localized orbitals and molecular symmetry. Consider for example an atomic system consisting of two electrons in an $(s)$ orbital and two electrons in a $(2px)$ orbital, both of which are self-consistent-field orbitals. Since they belong to irreducible representations of the atomic symmetry group, they are in fact the canonical orbitals of this system. Let these two self-consistent-field orbitals be denoted by $(s)$ and $(2p)$, and let $(h_+)$ and $(h_-)$ denote the two *digonal hybrid orbitals* defined by

$$(h_+) = [(s) + (2p)]/\sqrt{2}, \ (h_-) = [(s) - (2p)]/\sqrt{2}, \tag{37}$$

which point in opposite directions. It can be readily seen that the $\{(s), (2p)\}$ set, as well as the $\{(h_+), (h_-)\}$ set, *both* separately satisfy the localization Eqs. (33), which indicates that one of these two sets is minimally localized and the other one is maximally localized. A closer quantitative examination shows that the $(s)$, $(2p)$ set is maximally localized if the $(s)$ orbital is a $(1s)$ orbital, whereas the $(h_+)$, $(h_-)$ set is maximally localized if the $(s)$ orbital is a $(2s)$ orbital. More generally, if the $(s)$ orbital is *much* smaller in diameter than the $(p)$ orbital, or vice versa, then these two symmetry orbitals are more separate from each other than their hybrids, whereas, if both orbitals have about the same diameter, the reverse is the case.[27] Thus, under appropriate conditions the localized orbitals may be canonical and, hence, symmetry orbitals. Under other conditions, they are orbitals, like $(h_+)$ and $(h_-)$, which are exactly *identical in shape,* and which are *permuted among each other* by certain operations of the symmetry group. In the latter case, the localized orbitals are called *equivalent orbitals.*[28]

Of considerable interest is the case when the occupied canonical orbitals are the three atomic orbitals $(2s)$, $(2px)$, $(2py)$, as well as the case when the occupied canonical orbitals are the four atomic orbitals $(2s)$, $(2px)$, $(2py)$,

($2pz$). In the former case, the corresponding localized orbitals are the three well known orthogonal *trigonal* ($sp^2$) *hybrids*

$$t_k = \sqrt{1/3}\,(2s) + \sqrt{2/3}\,(2p_k)\ ,\ k = 1, 2, 3,$$

where ($2p_1$), ($2p_2$), ($2p_3$) are three $2p$ orbitals pointing to the corners of an equilateral triangle. In their most general form, these are given by

$$p_1 = \qquad\qquad\qquad \cos\alpha\,(2px) + \qquad\qquad\qquad\qquad \sin\alpha\,(2py),$$
$$p_2 = [-(1/2)\cos\alpha - (\sqrt{3}/2)\sin\alpha]\,(2px) + [(\sqrt{3}/2)\cos\alpha - (1/2)\sin\alpha)]\,(2py),$$
$$p_3 = [-(1/2)\cos\alpha + (\sqrt{3}/2)\sin\alpha]\,(2px) - [(\sqrt{3}/2)\cos\alpha + (1/2)\sin\alpha)]\,(2py),$$

where $\alpha$ is arbitrary and describes a rotation of the set of three hybrids around the $z$-axis. The trigonal hybrids are another example of the case where the localized orbitals are equivalent orbitals.[29] It may be noted that the full spherical symmetry is not used here, but only the rotational symmetry around the $z$-axis. For this reason we will encounter equivalent orbitals of this type in diatomic molecules also.

In the case that the occupied canonical orbitals are ($2s$), ($2px$), ($2py$), ($2pz$), the localized orbitals are given by the orthogonal *tetrahedral ($sp^3$) hybrids,* which point to the corners of a tetrahedron, a possible choice being

$$th_1 = (1/2)\,\{2s + 2px + 2py + 2pz\},$$
$$th_2 = (1/2)\,\{2s + 2px - 2py - 2pz\},$$
$$th_3 = (1/2)\,\{2s - 2px + 2py - 2pz\},$$
$$th_4 = (1/2)\,\{2s - 2px - 2py + 2pz\}.$$

Again they are equivalent orbitals.[30]

In molecules, equivalent orbitals can be observed, for example, in $CH_4$ and in $H_2O$. The ground state of $CH_4$ has four canonical valence orbitals belonging to the representations $A_1$ and $T_2$ respectively. The localized orbitals are oriented along the four bonds and their character is analogous to that of the aforementioned tetrahedral hybrids.[31] In $H_2O$ there are two bonding orbitals. In the canonical representation both of the bonding orbitals involve the oxygen atom and *both* hydrogen atoms. One is symmetric, the other antisymmetric with respect to the plane bisecting the molecule. In the localized representation, one finds two equivalent bonding orbitals, one concentrated around one OH bond, the other concentrated around the other OH bond.[32]

Finally, it must be mentioned that localized orbitals are not always simply related to symmetry. There are cases where the localized orbitals form neither a set of symmetry adapted orbitals, belonging to irreducible representations, nor a set of equivalent orbitals, permuting under symmetry operations, but a set of orbitals with little or no apparent relationship to the molecular symmetry group. This can occur, for example, when the symmetry is such that several sets of equivalent orbitals can be formed, for which a comparable degree of localization can be expected, but whose geometrical shapes conflict. [33]

# 5. Localized Orbitals in Diatomic Molecules

## 5.1. Correspondence between Canonical and Localized Orbital Structures

Usually the electronic structure of diatomic molecules is discussed in terms of the canonical molecular orbitals. In the case of homonuclear diatomics formed from atoms of the second period, these are the symmetry orbitals: $1\sigma_g$, $1\sigma_u$, $2\sigma_g$, $2\sigma_u$, $3\sigma_g$, $3\sigma_u$, $1\pi_u$, and $1\pi_g$. Contour diagrams for a set of such orbitals are shown in Fig. 1 (see p. 70). The contours are drawn in a plane containing the internuclear axis, which is also indicated. For $\sigma$ orbitals, three-dimensional contour-*surfaces* are obtained by spinning the contours around the internuclear axis. For $\pi$ orbitals the contours on a cross-section perpendicular to this axis are similar to those of atomic $p$ orbitals. *The diagrams represent contour lines of orbitals, not densities.* Where the orbitals are positive, the contours are drawn as solid lines; where the orbitals are negative, the contours are drawn as dashed lines; the nodes are drawn as dotted lines. In this figure, as well in subsequent ones, the outermost contour is chosen to approximate what can be thought of as the Van der Waal's radius of the orbitals. The increment in the wavefunction value, going from one contour to the next, is constant for each orbital plot. The increments are chosen by two considerations: On the one hand, we endeavored to place about ten contours between the outermost contour and the maximum; on the other hand, we have tried to use the same increment in as many different molecules as possible consistent with this goal. In Fig. 1, the outermost contour corresponds to $\pm 0.025$ Bohr$^{-3/2}$. The increment is $0.05$ Bohr$^{-3/2}$ for the valence shells and $0.2$ Bohr$^{-3/2}$ for the inner shells. For the latter the maximum (at the nucleus) in $\pm 7.165$ Bohr$^{-3/2}$, but no contours have been drawn beyond $\pm 2.025$ Bohr$^{-3/2}$. Corresponding contours are also omitted for the inner lobes of the $2\sigma_g$ and $2\sigma_u$ MO's, since they get too dense. The scale indicated in this and the subsequent figures is in Bohr radii.

The orbitals in Fig. 1 are those of the $F_2$ molecule. [34] However, different homonuclear molecules differ in the *overall scale* only, the *shapes* of the canonical orbitals are virtually identical for all of them.

In contrast, the forms of the corresponding localized orbitals depend characteristically upon the number of canonical orbitals which are occupied in any particular case. We discuss here, by way of an introduction, some examples of singlet states which are illustrated in Fig. 2 (see p. 71). The left hand column contains schematic diagrams of the canonical orbitals in the valence shell. Of the two $1\pi_u$ orbitals only one is indicated, and similarly with the $1\pi_g$ orbital.

Each of the remaining columns contains schematic sketches of the set of localized orbitals which results when certain of the canonical orbitals are occupied. Only the valence orbitals are shown; the inner shells are approximately atomic ($1s$) orbitals. Thus the fourth column shows, e.g., the three localized orbitals which are obtained when the canonical orbitals $2\sigma_g$, $2\sigma_u$, and $3\sigma_g$ are occupied.

In the case of two valence electrons there is hardly any difference between the localized orbital and the canonical valence orbital, except for the fact that the localization has separated the valence shell somewhat from the other shells. — In the case of four valence electrons, the sigma bonding and the sigma antibonding canonical orbitals yield two equivalent localized orbitals which resemble distorted atomic ($2s$) orbitals on each of the two atoms. They are precursors of what will be seen to be sigma lone pairs and are denoted by $\sigma l$ and $\sigma l'$. The absence of a bond can be ascribed to the nonbonded repulsion between these orbitals. This corresponds to the case of the unstable $Be_2$ molecule. — For six valence electrons, the case illustrated is that occurring when the canonical orbitals $2\sigma_g$, $2\sigma_u$, and $3\sigma_g$ are doubly occupied. This occupancy is seen to correspond to a localized structure consisting of a bonding orbital, denoted by a $\sigma b$, and two lone pair orbitals, $\sigma l$ and $\sigma l'$, one on each atom. Although the lone pairs have marked ($p$) character, they still contain more ($s$) than ($p$) character. The bonding localized orbitals, in contrast, contain more ($p$) than ($s$) character. This electronic structure corresponds to a low excited state of the $B_2$ molecule. — The next column illustrates the ten valence electron case, corresponding to the $N_2$ molecule. All canonical orbitals up to the two $1\pi_u$ orbitals are occupied. The localized structure has one lone pair on each of the two atoms, denoted by $\sigma l$ and $\sigma l'$, and three bonding orbitals which are arranged in a trigonally symmetric fashion around the bond axis between the two atoms and are commonly referred to as banana bonds. Only one of these is shown in the figure. The final column illustrates the case of 14 valence electrons, where all of the canonical orbitals up to $1\pi_g$ are doubly occupied, as is the case, for example, in the $F_2$ molecule. In this case the localized orbitals are a single sigma bond, denoted by $\sigma b$, and six lone pair orbitals, three on each atom. The three lone pair orbitals on any one atom form a set of three equivalent orbitals, denoted $t l_1$, $t l_2$, $t l_3$, which are arranged in a trigonally symmetric fashion. Only one of these is shown on each atom. As mentioned earlier, there is no preferred configuration for the lone pairs $t l_1$, $t l_2$, $t l_3$ relative to the lone pairs $t l_1'$, $t l_2'$, $t l_3'$.

The trigonal bond orbitals in the ten valence electron system as well as the two sets of trigonal lone pair orbitals in the 14 valence electron system are superpositions of $\pi$ orbitals and $\sigma$ orbitals. The formation of such trigonally symmetric molecular orbitals from $\sigma$-type and $\pi$-type molecular orbitals is entirely analogous in character to the formation of the three ($sp^2$) hybrid atomic orbitals from one ($s$) and two ($p$) atomic orbitals which was discussed in the preceding section. This can be visualized by looking at the diatomic molecule

49

along the internuclear axis. In a plane perpendicular to the axis contours of the canonical (localized) molecular orbitals look very much like contours of the canonical (hybridized) atomic orbitals. It is therefore apparent that an SCF wave function which, on the basis of its canonical orbitals, can be said to describe a sigma bond and two $\pi$ bonds, can equally well be said to describe three trigonally equivalent banana bonds, if one considers its representation in terms of localized orbitals. The two descriptions are quantum mechanically equivalent and any debate about the relative merits of one versus the other is therefore entirely void of substance within the self-consistent-field approximation.

In the following we shall discuss these localized orbital structures and those of some heteronuclear diatomics on the basis of accurate diagrams which were obtained from minimal basis set *ab initio* calculations. The numerical results on which these diagrams are based have been reported elsewhere. [35]

### 5.2. Sigma Bonds

Fig. 3 (see p. 72/73) exhibits contour diagrams of all localized molecular orbitals in the molecules $Li_2$ and LiH. [36] In $Li_2$ there are two inner shell orbitals and a bonding orbital. For LiH there are an inner shell orbital on lithium and a LiH bonding orbital. For the bonding orbital of $Li_2$, the outermost contour line corresponds to an orbital value of 0.005, the next contour line to 0.015, the next to 0.025 as indicated. Thus the increment is 0.01 $Bohr^{-3/2}$ in this case. By contrast, the outermost contour in the bonding orbital of LiH corresponds to a wavefunction value of 0.025 $Bohr^{-3/2}$, and the increment of the wavefunction value from one contour line to another ist also 0.025 $Bohr^{-3/2}$ in this case. The comparison of $Li_2$ and LiH shows that the $Li_2$ valence orbital is considerably larger than the LiH valence orbital and, moreover, that its maximum is much lower. In short, it is a much less compact orbital. We also see that the bonding orbitals of $Li_2$ and LiH have a rather strong negative peak near the Li nucleus which establishes orthogonality to the inner shells.

For the inner shells the outermost contour is again 0.025 $Bohr^{-3/2}$. They are much steeper and, therefore, the increment is here 0.2 $Bohr^{-3/2}$. Three of these inner shell contours are drawn. If the remaining inner shell contours were drawn, the inner part would be solid black. For this reason, the inner shell contours are not drawn beyond the third one and, instead, the value of the inner shell orbital at the position of the nucleus has been written into the diagram. From the figure, it is obvious that the inner shell of lithium is very similar in $Li_2$ and LiH, and in a very practical sense transferable. However, note that the localized inner shell orbital of the lithium atom has a slight negative tail towards the other atom which yields a very small amount of antibinding.

## 5.3. Sigma Bonds and Sigma Lone Pairs

Fig. 4 (see p. 74/75) shows all localized orbitals for the ground state of the BH molecule and the $^1\Sigma_g^+$ excited state of $B_2$.[37] These are again rotationally symmetric orbitals, i.e., sigma type orbitals, and the complete contour surfaces can be obtained by spinning around the indicated axis. In all orbitals shown the outermost contour corresponds to a wavefunction value of 0.025 Bohr$^{-3/2}$. For all valence shell orbitals the increment from one contour to another is 0.025 Bohr$^{-3/2}$. For the inner shells the increment is again 0.2 Bohr$^{-3/2}$, but only three contours and the wavefunction values at the nuclear positions are shown.

From the orbital distribution it is seen that the lone pair orbitals have almost all their density on that side of the atom which points away from the bond, whereas the bonding orbital has almost all its density in between the two atoms. There is of course some local overlap between the orbitals; in particular, the bonding orbital has some negative contributions in the lone pair region and the lone pair orbital has some negative contribution in the bonding region, so that the resulting orbitals will be orthogonal to each other. It is evident that the positive contours of the orbital have very similar distributions in $B_2$ and BH, as one would like to see them have. It is gratifying that the negative sides are only somewhat different even though rather different atoms are involved.

For the inner shell orbitals, too, one finds near-perfect transferability as was the case for lithium.

## 5.4. Sigma Bonds and Triple Lone Pairs

Fig. 5 (see p. 76/77) exhibits the localized orbital structure of the $F_2$ molecule and that of the FH molecule. [38] As was discussed earlier, the $F_2$ molecule consists of one localized orbital representing a single sigma bond and six lone pair orbitals, three on each atom, which accommodate the twelve lone pair electrons. All orbitals are much more contracted than those of boron, because of the higher nuclear charge of fluorine (note that the scale of all figures is the same). The outermost contour corresponds again to 0.025 Bohr$^{-3/2}$, but the increment between adjacent contours in the valence shell is now 0.05 Bohr$^{-3/2}$ because of the greater compactness of the orbitals. The contour surfaces of the bonding orbital are again obtained by spinning it around the nuclear axis.

For the lone pair orbitals the situation is somewhat more complicated. As mentioned before there are three trigonally equivalent lone pair orbitals at each end of the molecule which are arranged at 120° to each other, only one of which is shown on each atom. It can be observed that the lone pair orbital looks very much like a (s–p) hybrid on that particular atom, except for the little appendage which reaches over to the other atom. By connecting the position of the nucleus with the maximum of the lone pair orbital one can define an approximate axis of the lone pair orbital. The three dimensional contours of this lone pair are approximated by spinning the orbital around this axis, except in the region near the other atom. It is of interest that this axis of the lone pair orbital is not very far from being perpendicular to the internuclear axis. It is much less inclined toward the back of the molecule than it would be in the case of tetrahedral hybridization. The reason for this is that the repulsion between the three lone pair orbitals is stronger in its effect than the repulsion between any one lone pair and the bonding orbital. This has to do with the fact that the lone pair orbitals have more $(s)$ character than the bonding orbital, and is in agreement with Gillespie's previously postulated model. [39]

The relation between $F_2$ and FH is similar to that observed between $B_2$ and BH. The FH molecule has a sigma bonding orbital and has three trigonally equivalent lone pairs which are almost identical in character and shape to the corresponding lone pairs of $F_2$. These contracted lone pairs are less sensitive to the other atom than those on B. We also find nearly complete transferability between the inner shells. Here again the outermost contour is 0.025 Bohr$^{-3/2}$ and the increment of those contours which are shown is 0.2 Bohr$^{-3/2}$.

The main difference between the two molecules lies in the bonding orbital. It may however be noted that the part of the bonding orbital near the fluorine nucleus is rather similar in the two systems. In both molecules the bonding orbital exhibits a maximum close to the fluorine atom which arises from the increased $(2p\sigma)$ admixture to the bonding orbital. Thus, proceeding from F along the internuclear axis the orbital rises from the value zero, at the atom, to the maximum, and then begins to drop in the bond region. This is different from what was seen in $B_2$ and BH.

### 5.5. Triple Bond and Sigma Lone Pairs

The left side of Fig. 6 (see p. 78/79) shows the localized orbital structure of the $N_2$ molecule. [40] As mentioned above, we have here one lone pair on each nitrogen atom and three trigonally equivalent banana bonds between the two atoms. The outermost contour in each orbital shown in this figure is again 0.025 Bohr$^{-3/2}$. The increment is 0.05 Bohr$^{-3/2}$ for the valence orbitals and 0.2 Bohr$^{-3/2}$ for the inner shell orbitals. There are three bonding orbitals which are arranged in a trigonally symmetric fashion around the internuclear axis; only one of them is shown in the figure. For this one, the contour lines in the plane containing the orbital maximum and the inter nuclear axis are exhibited. The three dimensional contours can be expected to form a three dimensional cloud essentially above the internuclear axis. The cross section in a plane perpendicular to the axis should be roughly that of a

$sp^2$ hybrid. A distinct maximum is observed near each nucleus, but it is less pronounced than those seen in the $op$ orbital of $F_2$ and HF.

The right side of the figure shows the localized structure of the CO molecule. The quantitative meanings of the contours are the same as in $N_2$ [41]. The CO molecule is isoelectronic with $N_2$, and the localized orbital structure brings this out very clearly. One can imagine the CO structure obtained from the $N_2$ structure by transferring one proton charge from the left nucleus to the right nucleus. This results in the contraction of the lone pair near the O nucleus and the expansion of the lone pair near the C nucleus as compared to the nitrogen situation. Due to the orthogonality requirement, the negative contours of the carbon lone pair are less spread toward oxygen than are the negative contours of the oxygen lone pair toward carbon. The negative contours of the $N_2$ lone pairs are intermediate in spread. Moreover, each of the three bonding orbitals is polarized towards the oxygen atom. Finally the inner shell of oxygen is smaller than that of nitrogen, whereas that of carbon is bigger.

The third molecule in this isoelectronic series, BF, is shown on the left hand side of Fig. 7 (see p. 80/81). [42] The localized orbitals are completely analogous to $N_2$ and CO, except that the charge difference between B and F is even greater than that between C and O. Hence the lone pair of fluorine is even more contracted near the F nucleus and more diffuse toward the B nucleus, whereas the lone pair of boron is more expanded near the B nucleus and less spread toward the F nucleus. The inner shell of fluorine is also contracted; the inner shell of boron is expanded. The three trigonal bonding orbitals are even more polarized towards the heavy atom than they were in CO and concomitantly acquire more fluorine character. In fact, near the fluorine atom the trigonal bonding orbitals look similar to the trigonal lone pairs of fluorine found in $F_2$ and FH, except that the axis is, of course, tilted towards the bond. Since the boron lone pair orbital is considerably more extended, the increment between adjacent contours is chosen to be 0.025 $Bohr^{-3/2}$, that is, a step by two contours in the B lone pair corresponds to a step by one contour in the F lone pair or the bonding orbital in this figure.

The right hand side of Fig. 7 (see p. 80/81) shows the LiF molecule. [43] Although it is not isoelectronic with BF, its localized structure is not so different because it can be thought of as being obtained from the BF molecule by removing two positive nuclear charges and the two lone pair electrons from the boron atom. There remain then the fluorine lone pair and inner shell orbitals, all of which are similar to those found in BF, and the trigonal bonding orbitals which, although they are even more polarized towards the fluorine atom, still show some similarity to those found in BF. The inner shell in lithium is of course considerably larger, and similar to that found in $Li_2$ and in LiH.

## 5.6. Triple Bond and No Lone Pair

The ground state of the NH molecule has the electron configuration $^3\Sigma (1\sigma)^2$ $(2\sigma)^2$ $(3\sigma)^2$ $(\pi x)$ $(\pi y)$. When the $\pi x$, $\pi y$ orbitals are excluded from the localization procedure, the localized structure consists of an inner shell on nitrogen, a lone pair on nitrogen and a sigma bonding orbital. A visualization of this can be obtained from the oxygen atom in the electron configuration $(1s)^2$ $(2s)^2$ $(2p\sigma)^2$ $(2p\pi)$ $(2p\bar{\pi})$. First, we hybridize the $(2s)$ and $(2p\sigma)$ orbitals to obtain digonal hybrids. Then, we imagine removing a proton from inside the O nucleus to obtain N and H nuclei. The digonal hybrids on O then become a lone pair on N and a $\sigma$ bonding orbital.

In Fig. 8 (see p. 82), there is shown the localized orbital structure of the $^1\Sigma$ $(1\sigma)^2$ $(2\sigma)^2$ $(1\pi)^4$ excited state, which can be thought to result by promoting two electrons from the $\sigma$ lone pair into the non-bonding orbitals $(\pi x)$ and $(\pi y)$, [44] which are essentially atomic $(p)$ orbitals. When this structure is localized, the sigma bonding orbital combines with the $\pi$ orbitals to form three trigonally arranged banana bonds between the nitrogen and the hydrogen, only one of which is shown in the figure. Unlike other cases involving a common atom in different molecules (e.g. $B_2$, BH, BF), the inner-shell in NH is more spherically symmetric about N than are the inner shells in $N_2$, i.e. the atomic $(1s)$ orbitals in $N_2$ are mixed with the valence atomic orbitals to a slightly greater extent than the nitrogen $(1s)$ orbital in NH. Perhaps, this is due to the fact that there is no longer a sigma lone pair. The resulting structure of bonding orbitals is analogous to that found in LiF. This example shows how localization can lead to different localized orbitals in different states of a molecule.

## 5.7. Comparison of Sigma Lone Pair Orbitals in Different Molecules

In Fig. 9 (see p. 82/83) we have collected all sigma lone pairs which were discussed in the molecules considered. They are arranged according to increasing nuclear charge. The overall impression is that of a great similarity in the geometrical shapes of the lone pair orbitals. In all cases the density is concentrated on that side of the atom which is away from the bond, and in all cases the shape is that of an $s$–$p$ hybrid with considerable $(s)$ character. Except for F, the latter is always larger than 50%. The larger the fraction of the valence orbitals which are lone pairs, the larger the $(2s)$ character of the lone pair orbitals. [45] In all cases, there is a smaller negative contribution towards the second atom. Even though different atoms are involved, the general shape of this usually weak antibonding contribution is fairly uniform. The general lone pair shape is preserved throughout the whole series, even though the overall size of the lone pair orbital decreases progressively as one proceeds from lighter to heavier nuclei.

All lone pair orbitals have a node between the two atoms and, hence, have a slightly antibonding character. This destabilizing effect of the lone pair localized molecular orbitals corresponds to the nonbonded repulsions *between* lone pair atomic orbitals in the valence bond theory. In the MO theory all bonding and antibonding resonance effects can be described as sums of contributions from *orthogonal* molecular orbitals. Hence, the "nonbonded repulsions" appear here as *"intra-orbital"* antibonding effects in contrast to the valence-bond description.

Very close transferability can be observed between the three boron and the two fluorine lone pair orbitals. From these results, it appears virtually certain that, if one has a localized orbital in a larger molecule, and if one changes some of the atoms which the orbital itself does not reach, then almost absolute transferability can be expected. We are currently investigating such cases.

## 5.8. Comparison of Sigma Bonding Orbitals in Different Molecules

All sigma bonding orbitals which were encountered in the molecules considered are collected in Fig. 10 (see p. 84/85). To save space, the outer part of the $Li_2$ molecule has been removed (c.f. Fig. 3, p. 72/73). The bonding orbitals show the overall contraction going from light atoms to heavy atoms. Also observe that in $B_2$ and $F_2$, the bonding orbital has negative parts in the lone pair regions, because it has to be orthogonal to the lone pairs, which is not the case in $Li_2$.

As regards the hydrides, it is of interest to compare the bonding orbitals of BH and FH with the corresponding lone pairs on B and F shown in Fig. 9 (see p. 82/83). The similarity in the overall size of the bonding and the lone pair orbitals is quite remarkable. This indicates that there must be a large degree of overlap between the H orbital and the $(sp)$ hybrid of the heavy atom contributing to the bonding orbital. However, it is apparent that this hybrid has more $(p)$ character than the lone pair. In going from LiH to BH to FH the bonding orbital acquires an increasingly greater $(p\sigma)$ character because of an increasing amount of non-bonded repulsion from lone pair electrons. This is manifest in the bonding orbital as an elongation and an increased number of negative contours outside the bond region from LiH to BH to FH.

## 5.9. Comparison of Trigonal Orbitals in Different Molecules

Fig. 11 (see p. 86/87) contains all trigonal orbitals which were encountered in the molecules considered. The bonding orbitals, in the left column, exhibit the increasing polarization from $N_2$ to LiF. Moreover, the inclination of the contributing $(sp)$ hybrid of the right atom into the bond region diminishes as the polarization increases, i.e., the axis of this hybrid is much closer to being perpendicular to the internuclear axis in LiF than in $N_2$. Clearly, an increase in $(p)$ character accompanies the diminshed inclination.

The lone pair hybrids in $F_2$ and FH are even more nearly perpendicular to the internuclear axis. They are very similar, but the one in FH is slightly more inclined away from the bond, i.e. it has a slightly lower ($p$) character. This is so, presumably, because the FH bonding orbital puts more charge in the immediate neighborhood of the F atom.

# 6. Localized Pi-Molecular Orbitals in Aromatic Hydrocarbons

## 6.1. Localization of Delocalized Electrons

In a crystal lattice where each atom contributes one atomic orbital, and where these orbitals are related to each other by the translations characteristic of the lattice, the molecular orbitals must belong to irreducible representations of the group of these translations and hence form so-called Bloch orbitals. [46] In many respects the Bloch orbitals have the character of plane waves which extend through the whole crystal and differ from each other by the number of nodes. Since a similar situation exists for the $\pi$ electrons in a planar aromatic system, its canonical orbitals are, in some ways, similar to Bloch orbitals. Here too, each atom contributes one atomic orbital, a $\pi$ orbital which is perpendicular to the plane of the molecule, and these $\pi$ orbitals are related to each other by translations along the bond skeleton. The bond skeleton forms a one dimensional, usually multiconnected grid and its invariance group is, of course, more complicated than a simple translational group. Nevertheless the canonical $\pi$ orbitals are well known to have characteristics similar to those of plane waves moving along the bond skeleton. In fact a detailed equivalence has been proved to exist between two simple approximations to these $\pi$-electron orbitals, namely the Hückel-Wheland approximation [47] and the free electron approximation. [48] All $\pi$ orbitals together form a set of molecular orbitals analogous to what in a crystal is called a Bloch band. The number of occupied $\pi$-molecular orbitals, usually about half of what would be a band in the crystal, corresponds to the Bloch states below the Fermi level. The Bloch orbitals, as well as the aromatic $\pi$ orbitals or the free electron orbitals, are said to be delocalized, because each of them extends over the whole molecular or crystal framework. On the basis of these delocalized orbitals, it is possible to understand the mobility of the electrons which, in a metal, gives rise to such effects as the electric conductivity, and, in an aromatic system, causes characteristic chemical behavior. In particular, the delocalization of the $\pi$ electrons has been recognized as the source of the additional stabilization of aromatic ring systems as compared to the stability which would be expected from a set of single and double bonds.

It is certainly possible to apply the localization procedure to the determinantal wave functions of the $\pi$ electrons and, thus, represent this system in terms of a set of occupied localized orbitals. (In view of the analogy between canonical $\pi$ orbitals and Bloch orbitals, it may be noted that the localized $\pi$ orbitals are not analogous to Wannier functions. [49] The latter are linear combinations of *all* Bloch orbitals and correspond to orthogonalized $\pi$-atomic orbitals.) One might wonder if anything interesting will be gained by looking from a localized standpoint at a system which is known as being delocalized. We shall see that this is indeed the case and that, in point of fact, the concept of delocalization itself will be clarified if looked at in terms of a localized formulation.

One might also wonder about the appropriateness of localizing the $\pi$-electron system alone, exclusive of the sigma electrons. Such localization means in effect that, among all orthogonal transformations of the type of Eq. (5), only a certain subclass is considered, namely those which consist of two independent blocks, each of which represents a smaller orthogonal transformation: the first block transforms only the $\pi$ orbitals, and the second transforms all other orbitals with no mixing admitted between the two types. Thus, one does not expect maximal localization, of course; but, on the other hand, this approach yields certain interesting information about the $\pi$ system by itself which is obscured if the $\sigma$ system is included.

## 6.2. Approximation to Canonical Orbitals

Since we are essentially interested in qualitative features, we choose the *canonical $\pi$* orbitals to be the *Hückel-Wheland* orbitals. [50] These are $\pi$-molecular orbitals which are expressed as linear combinations of the $(2pz)$ atomic orbitals on the various carbon atoms. If the latter are denoted by $p_1, p_2, p_3 \ldots$ then their overlap matrix is assumed to be [51]

$$(p_i | p_k) = \delta_{ik} + (1/4)M_{ik} \tag{38}$$

where the matrix elements $M_{ik}$, known as the Hückel or topological matrix, are unity if the index pair $(ik)$ denotes a pair of neighbor atoms and zero otherwise. That is to say, overlap is taken into account between neighbors but neglected between non-neighbors. The canonical orbitals are eigenfunctions of an effective one-electron hamiltonian, whose matrix likewise contains only neighbor interactions and is given by

$$(p_i | \mathcal{H} | p_k) = \alpha \delta_{ik} + \beta M_{ik}. \tag{39}$$

The Hückel-Wheland molecular orbitals are

$$\phi_n(x) = (1 + \mu_n/4)^{-1/2} \sum_k p_k(x) C_{kn}, \tag{40}$$

where the $\{C_{kn}\}$ are the eigenvectors and the $\mu_n$ are the eigenvalues of the Hückel matrix $M_{ik}$. [52] If the neighbor overlap $S$ is set equal to zero instead of 1/4, these orbitals become the well-known Hückel orbitals. In other words, the Hückel-Wheland orbitals are generalized from the Hückel orbitals to include overlap between neighbors.

In the spirit of the Hückel-Wheland theory, the Mulliken approximation for electron repulsion integrals [53] is used in the maximization of the localization sum (Eq. (28)). The details of the calculations are reported elsewhere. [54]

### 6.3. Localized $\pi$ Orbitals in Benzene

Fig. 12 (see p. 88 ) exhibits contour diagrams for localized $\pi$-MO's in benzene. The contours drawn in this and all subsequent figures represent values of the MO's in a plane which is parallel to the molecular plane and contains the maxima of all ($2pz$) atomic orbitals. Since the latter are chosen to have the orbital exponent 1.61789 in atomic units, the contour plot plane is the plane $z$ = = 0.61809 Bohr. Of course, in the molecular plane itself, all $\pi$-MO's have the value zero. The contours of each MO are obtained by dividing the difference between the maximum and the minimum function value into equal increments. In all cases, the function values are about 0.239 Bohr$^{-3/2}$ for the maximum contour, about $-0.072$ Bohr$^{-3/2}$ for the minimum contour, and about 0.016 Bohr$^{-3/2}$ for the increment.

The $\pi$-electronic system in benzene is interesting in that there exists an infinity of sets of localized $\pi$ orbitals with an equal degree of localization. The last row of Fig. 12 (see p. 88) exhibits one of these sets. The second, third and fourth diagrams of the row contain the contours of the three localized orbitals. It is seen that each orbital is essentially concentrated on two atoms and in the bond region between them, but it extends a little bit over two further adjacent atoms and has a slight negative contribution on the remaining two atoms. It is apparent that such a localized molecular orbital comes close to describing a double bond between the two atoms which contain the principal contributions. Thus the three localized orbitals in the last row can be said to represent a Kekulé type structure of the $\pi$-electronic system. [55] They are seen to be *equivalent* MO's. The first picture in the last row is a composite of the three localized orbitals just discussed. It is obtained by superimposing the fifth strongest contour of each of the three orbitals onto the bond skeleton. Thus each of the closed loops represents one of the three localized orbitals whose explicit contour diagram appears to the rigth in this row.

The first row represents an alternative set of equivalent localized orbitals which is as strongly localized as the one just discussed. [56] They extend essentially over three atoms. Whereas the Kekulé type localized orbitals are symmetric with respect to the plane bisecting a bond, the localized orbitals in the first row are symmetric with respect to a plane containing two opposite atoms. The negative lobe extends only over one atom, and the node passes almost exactly through the two remaining atoms. The composite picture of these localized orbitals, again obtained by drawing the fifth strongest contour into the benzene ring, is given as the first picture of the row. Since each of the contours covers three atoms, they overlap.

In the second and third rows there are given two other possible sets of localized orbitals *with equal degrees* of localization. They can be condidered as intermediate between those of the first row and those of the last row. The orbitals of the second row can be hypothetically obtained from those of the first row by deforming the orbital around the benzene ring in a clockwise direction. If the orbital is moved even further in that direction, one can pass from the second row to the third row, and eventually from the third row to the fourth row. In fact, this transition from the first row to the last row is a continuous process, and there exist infinitely many sets of localized orbitals of intermediate character, only two of which have been indicated in the second and the third rows. [57] Again the first column contains the superimposed fifth strongest contours for each set of molecular orbitals (in the case of the orbitals of the third row the fifth strongest contour happens to divide up into two disconnected parts).

We have discussed these contours of benzene in some detail since they will help us understand localized orbitals which appear in other aromatic systems.

## 6.4. Classification of Localized $\pi$ Orbitals

In view of the ambiguity of the localized orbitals in benzene and also the fact that, for any of the larger aromatic ring systems, there exist a considerable number of classical Kekulé structures, it would not be surprising to find several sets of equally or near equally localized $\pi$ orbitals for any or some of these $\pi$-electronic systems. [58] However, the investigation of 21 aromatic ring systems, some of them quite large, showed that in each of them *only one set of localized orbitals existed,* even though considerable effort was made to search for alternative sets. [54]

An examination of these systems revealed that certain types of localized orbitals occurred over and over, so a classification of all occurring localized $\pi$ orbitals into a small number of different types proved possible. This classification is embodied in Fig. 13 (see p. 89). There are four "pure" types: the orbitals denoted by $\pi\ell2$, $\pi\ell3$, $\pi\ell4$, and $\pi\ell2'$. The types $\pi\ell2$ and $\pi\ell3$ are the two symmetric types found in the benzene molecule. The two intermediate types denoted by $\pi\ell23$ were also found in benzene. The type $\pi\ell4$ occurs in aromatic molecules which have a joint and extends over four atoms, the joint atom being at the center and carrying most of the charge. The type $\pi\ell2'$ can occur in molecules which have adjacent joint atoms and in which the main orbital density is in a joint-joint bond. It extends somewhat onto all four atoms nearest to this bond. An orbital of this type is found, for example, in naphthalene. Just as in benzene we found intermediate types between $\pi\ell2$ and $\pi\ell3$, there occur intermediate types of localized $\pi$ orbitals between $\pi\ell3$ and $\pi\ell4$, between $\pi\ell4$ and $\pi\ell2'$, and again between $\pi\ell2$ and $\pi\ell2'$. The possible intermediate type orbitals form a continuous sequence and for each category one or two typical examples are indicated. These intermediate examples are chosen in such a fashion that one can imagine the transition from one type to the other by a gradual distortion of the orbitals in the appropriate directions. As will be seen below, each of the localized orbitals indicated has negative lobes associated with it, but these are omitted from the diagram for clarity.

We shall now substantiate this classification by showing examples of each type of localized $\pi$ orbital in a number of different systems.

### 6.5. $\pi\ell2$-Type Localized Orbitals

Figs. 14, 15, 16, 17 (see pp. 90/91, 92/93, 94/95, 96/97) show examples of localized orbitals of type $\pi\ell2$ and $\pi\ell23$ in multi-ring systems. The orbitals whose contours are shown in Figs. 14 and 15 (see pp. 90/91, 92/93) are very symmetric about the two principal atoms, i.e. of type $\pi\ell2$. The orbitals shown in Fig. 16 and 17 (see pp. 94/95, 96/97) are of type $\pi\ell23$. In all cases shown, the localized orbitals are situated on an "outside" benzene ring, where there are four non-joint atoms.

In Figs. 18 and 19 (see pp. 98/99 and 100/101) we show localized $\pi$ MO's occurring on parts of a benzene ring which contains only two non-joint atoms. They are of the same basic type as those found on the branches containing four non-joint atoms. The orbitals in Fig. 18 (see p. 98/99) are symmetric or nearly so, i.e. of type $\pi\ell2$, whereas those in Fig. 19 (see p. 100/101) are somewhat asymmetric, i.e. type $\pi\ell23$. The contour plots shown in Figs. 14 to 19 (see pp. 90–101) exhibit a most remarkable similarity among the localized $\pi$ orbitals in many different aromatic hydrocarbons.

### 6.6. Transition from $\pi\ell2$ to $\pi\ell3$

In outer parts of benzene rings containing three non-joint atoms, one observes localized orbitals of type $\pi\ell2$ as well as as of type $\pi\ell3$ and in some instances intermediate types. This is illustrated in Fig. 20 (see p. 102/103) where, in the upper left hand corner, pyrene exemplifies a pure $\pi\ell3$-type orbital. As we go to the right in the first row and then into the second row, we see orbitals on similar three atom branches which become more and more asymmetric and finally are of type $\pi\ell2$. Proceeding through the second row into the third row, we have again a transition to type $\pi\ell3$, but now one of the three atoms is a joint atom.

60

## 6.7. Transition from $\pi\ell 3$ to $\pi\ell 4$

*Outside branch involving one non-joint atom*

In the next three figures, there are collected a number of localized $\pi$-MO's which are arranged in a sequence that illustrates the transition from type $\pi\ell 3$ to type $\pi\ell 4$. Starting with Fig. 21 (see p. 104/105) on the top and going down, we first see a number of orbitals of type $\pi\ell 3$. The lower three orbitals show however some deformation from right to left, and appear as the intermediate type $\pi\ell 34$. Continuing on top of Fig. 22 (see p. 106/107) further $\pi\ell 34$ orbitals follow and, proceeding down the figure, we gradually approach the type $\pi\ell 4$. The sequence continues on Fig. 23 (see p. 108/109) and leads to several orbitals which are almost pure $\pi\ell 4$. Note that the orbitals of type $\pi\ell 34$, as well as the orbitals of type $\pi\ell 4$, involve at least two joint atoms.

*Outside branch involving three non-joint atoms*

Fig. 24 (see p. 110/111) shows a sequence of localized $\pi$-MO's which illustrate a second transition to type $\pi\ell 4$. We start out with orbitals of type $\pi\ell 3$ localized on an outside branch containing three non-joint atoms. Proceeding across the first row and then the second row of the figure, a sequence of orbitals is seen which represents the transition from type $\pi\ell 3$ to type $\pi\ell 4$, as was discussed in the previous section for an outside branch involving one non-joint atom.

## 6.8. Transition from $\pi\ell 4$ to $\pi\ell 2'$

The last row of Fig. 24 (see p. 110/111) and the first two rows of Fig. 25 (see p. 112/113) exhibit a sequence of orbitals representing the gradual transition from type $\pi\ell 4$ to type $\pi\ell 2'$. The lower part of Fig. 25 (see p. 112/113) contains examples of almost pure type $\pi\ell 2'$, with slight asymmetry in the direction of $\pi\ell 4$.

## 6.9. Transition from $\pi\ell 2'$ to $\pi\ell 2$

The upper part of Fig. 26 (see p. 114/115) contains several examples of localized $\pi$ orbitals of pure type $\pi\ell 2'$, the prototype being naphthalene. It is seen that, even in fairly asymmetric molecular situations, the localized orbitals are still of quite pure type $\pi\ell 2'$. Really strong asymmetry is seen in the two molecules at the bottom of the figure, which show orbitals of the type $\pi\ell 22'$. The type $\pi\ell 2$ would result from type $\pi\ell 2'$ if all the contours from the right side of the orbital were pushed over to the left side. On the last figure (see Fig. 27, p. 116/117), there are further examples of pure and deformed type $\pi\ell 2'$. It can be seen that in sufficiently asymmetric molecular situations this type of orbital, which extends over four "minor" atoms, can have quite irregular forms.

## 6.10. Localized Structures of Aromatic $\pi$ Systems

After having discussed the possible types of localized $\pi$-MO's which are encountered in $\pi$-electronic systems, we can now examine the total localized structures for the molecules investigated. Fig. 28 (see p. 118/119) contains a number of aromatic systems whose localized structure involves localized orbitals of type $\pi\ell2$ and $\pi\ell2'$. The diagrams are of the same nature as those in the first column of Fig. 12 (see p. 88), i.e. the fifth strongest contour of each localized orbital in the particular aromatic system is drawn on the molecular skeleton. The first molecule on this figure is naphthalene, which has four $\pi\ell2$ orbitals and, on the center bond, a $\pi\ell2'$ orbital. It may be noticed that the fifth strongest contour of the $\pi\ell2'$ orbital is only little different from those of the $\pi\ell2$ orbitals, but, from Fig. 26 (see p. 114/115) we know that the center bond is actually a $\pi\ell2'$ orbital. It is evident that this localized structure corresponds to the symmetric Kekulé structure of naphthalene. The other molecules shown on this figure are of a similar character. On the non-joint atoms we find orbitals of type $\pi\ell2$ whereas, on the joint atoms, we find orbitals of type $\pi\ell2'$. *In all cases, the arrangement of localized orbitals corresponds to a Kekulé structure where as many rings as possible have the benzenoid form, in agreement with Fries' rule.* [59] Note that although there are two resonance forms possible for 1, 2, 3, 4-dibenzanthracene and benzanthracene (the second and fourth molecules, respectively, in the right column of Fig. 28, p. 118/119) which satisfy Fries' rule, the observed localized orbital structures correspond to the ones in which the $\pi\ell2'$ orbitals are separated one from the other as much as possible.

Whereas the preceding figure showed a number of catacondensed systems exhibiting what might be called Kekulé-type localization, i.e. containing only orbitals of types $\pi\ell2$ and $\pi\ell2'$, Fig. 29 (see p. 120/121) contains a number of catacondensed systems where other localized $\pi$-orbitals occur. In the higher polyacenes, on the upper part of the figure, there are several ways to arrange a maximum number of benzenoid forms and the localized orbitals are "caught in a dilemma": On the two end branches, which contain four non-joint atoms, there result, as usual, two localized orbitals of type $\pi\ell2$. If the remaining localized orbitals were also of type $\pi\ell2$ and $\pi\ell2'$, we could obtain an asymmetric orbital structure similar to one of the asymmetric Kekulé structures. Instead, the localization procedure yields either of two identical sets of localized orbitals of type $\pi\ell3$ on one side of each molecule, and type $\pi\ell4$ and/ or $\pi\ell3$ on the other side. A somewhat similar situation exists in the two molecules at the bottom. Each can be considered as being made from two naphthalene type ends, which are joined together by an additional central ring. One sees that, on the naphthalene-type ends, orbitals of type $\pi\ell23$ and $\pi\ell2'$ occur and that, on the ring which joins these ends, three localized orbitals of type $\pi\ell3$, or perhaps $\pi\ell34$, achieve an optimal junction. The azulene molecule shown on the upper right is seen to have four orbitals of type $\pi\ell3$ and one of type $\pi\ell2$. Its localized orbitals are completely unrelated to the plane of symmetry going through the length of the molecule.

Fig. 30 (see p. 122/123) shows six pericondensed systems. The localized structures of the three molecules in the top row appear to be similar to those of Fig. 28 (see p. 118/119). This is however true only for the molecule on the upper right, coronene, whose outer branches consist of two non-joint atoms. In contrast, the first two molecules in the first row contain branches with three non-joint atoms and, on these, the $\pi\ell23$ localized orbitals are seen to be much like the $\pi\ell23$ we found on the naphthalenic ends of the dibenzanthracenes in the previous figure, but more nearly $\pi\ell3$ orbitals. Nevertheless, these structures are close to what would be predicted if perylene (to the left) were thought of as two non-interacting naphthalene fragments and benzoperylene (in the center) were thought of as the catacondensed skeleton

The molecules in the second row of this figure show localized orbitals of all types. The first on the lower left, pyrene, has type $\pi\ell2$ on the outer branches containing two non-joint atoms, type $\pi\ell3$ on the outer branches containing three non-joint atoms, and a type $\pi\ell4$ on one of the inner joint atoms. Benzpyrene, in the center of the bottom row, is similar to pyrene in its upper part and to naphthalene in its lower part. Correspondingly, the localized structure is in fact similar to that of pyrene in the upper part and similar to that of naphthalene in the lower part. In the center of the molecule, where these two partial systems are fused, a sort of compromise is reached by a suitable deformation of what was a pure $\pi\ell4$ orbital in pyrene. The molecule at the lower. right, anthanthrene, has a rather ingenious combination of interlocking localized orbitals of type $\pi\ell2$, $\pi\ell3$, $\pi\ell4$ to cover the condensed network.

## 6.11. Delocalization and Resonance Energy

Aromatic molecules are more stable than one would expect them to be, if they were considered as consisting of unrelated single and double bonds. The fact that their double bonds are "conjugated" leads to an additional stability which, especially for molecules consisting of six membered rings, is one of the properties of the aromatic character. A measure of this additional stabilization is the "experimental resonance energy", which is defined as the difference between the actual energy of formation of the molecule and the hypothetical energy of formation, obtained by adding up all single and double bond energies of one of the Kekulé structures, where for the double bond energy the ethylene value is taken.

On the theoretical side it has been observed that, in any aromatic system, the sum of the $\pi$-orbital energy contributions of the canonical Hückel-Wheland orbitals is always substantially lower than the energy obtained by multiplying the $\pi$-electron energy contribution of the Hückel-Wheland orbital of ethylene by the number of double bonds. Although the true $\pi$-electron energy is of course not the sum of the orbital energy contributions, it still appears reasonable to consider this behavior of the sum of the orbital-energy-contributions as being related to the experimentally observed aromatic stability. This difference in the orbital sum is called the "theoretical resonance energy."

It is commonly stated in the molecular orbital theory that the theoretical resonance energy is due to *the delocalization of the canonical $\pi$-electronic orbitals.* Now it is true that all Hückel-Wheland orbitals, like the Bloch orbitals or the free electron orbitals, do cover the total molecular skeleton, and that the lowest ones have low energies, since they have only a few nodes. The higher ones acquire, however, more and more nodes, and their orbital energy contributions increase correspondingly. For example in the systems considered here, the orbital energy contributions of the canonical orbitals vary between $(\alpha + 1.6\gamma)$ and $(\alpha + 0.2\gamma)$, where $\gamma$ is the resonance integral, given by

$$\gamma = \beta - \alpha S \approx -2 \text{eV}.$$

The orbital energy contribution of ethylene, on the other hand, is $(\alpha + 0.8\gamma)$. It is therefore unclear why the sum of the orbital energy contributions should always be lower than the sum of the corresponding number of ethylene energies. Furthermore, in view of the theory presented in the first section of this paper, it is true that molecular orbitals do not *have* to be chosen as being completely delocalized, and that the extreme delocalization of the Hückel-Wheland orbitals represents an arbitrary choice. In fact, if one wishes to assess the degree of *delocalization,* one is loading the dice in a confusing manner, if one bases the discussion on the *most* delocalized set of molecular orbitals available. However, *a meaningful answer to this question can be expected if one shows that it is not possible to localize the molecular orbitals in such systems to as high a degree as in systems containing single bonds only.* It is therefore not surprising that we can obtain a much better insight into the origin of the resonance energy if we use the localized orbital representation.

The orbital energy contributions of the canonical orbitals, $\epsilon_i$, are the eigenvalues of the Hückel-Wheland hamiltonian (39). If the localized orbitals $\lambda_K$ are given in terms of the canonical orbitals $\phi_i$ by

$$\lambda_K = \sum_{i=1}^{N} \phi_i T_{iK} \tag{41}$$

with $T$ defined as in Eq. (6), then the localized orbital contributions $\eta_K$ are defined to be

$$\eta_K = (\lambda_K |\mathcal{H}| \lambda_K) = \sum_{i=1}^{N} (T_{iK})^2 \epsilon_K. \qquad (42)$$

It is readily seen that the sum of the orbital contributions is invariant, i.e.

$$\sum_{i=1}^{N} \epsilon_i = \sum_{K=1}^{N} \eta_K. \qquad (43)$$

*The theoretical resonance energy can therefore be calculated either from the localized orbitals or from the canonical orbitals.*

Since all localized orbitals have approximately the same spatial extension, it stands to reason that their orbital energies $\eta_n$ should be of the same order of magnitude. In fact, one finds for the four main types of localized $\pi$ electron orbitals the following orbital energy values:

$\pi\ell 2$, on branch of two non-joint atoms: $\qquad \alpha + .90\gamma > \eta > \alpha + .96\gamma,$

$\pi\ell 2$, on branch of four non-joint atoms: $\Big\}$

$\pi\ell 3$, centered on non-joint atom: $\qquad \alpha + .92\gamma > \eta > \alpha + .98\gamma,$

$\pi\ell 2$, one atom a joint atom: $\qquad \alpha + 1.00\gamma > \eta > \alpha + 1.02\,\gamma,$

$\pi\ell 3$, centered on joint atom: $\qquad \alpha + 1.03\gamma > \eta > \alpha + 1.08\,\gamma,$

$\pi\ell 4$ and $\pi\ell 2'$: $\qquad \alpha + 1.08\gamma > \eta > \alpha + 1.12\,\gamma,$

which are obtained by transforming the Hückel-Wheland energies. We thus find that the *orbital energy contributions of all localized $\pi$-electronic orbitals are rather close to each other in energy, and considerably more negative than the orbital energy contribution of ethylene* ($\alpha + 0.8\gamma$).

The reason for this becomes apparent when one compares the shapes of the localized $\pi$ orbitals with that of the ethylene $\pi$ orbital. All of the former have a positive lobe which extends over *at least three* atoms. In contrast, the ethylene orbital is strictly limited to *two* atoms, i.e., the ethylene $\pi$ orbital is considerably more localized than even the maximally localized orbitals occurring in the aromatic systems. This, then, is the origin of the theoretical resonance energy: *the additional stabilization that is found in aromatic conjugated systems arises from the fact that even the maximally localized $\pi$ orbitals are still more delocalized than the ethylene orbital.* The localized description permits us therefore to be more precise and suggests that resonance stabilization in aromatic molecules be ascribed to a *"local delocalization" of each localized orbital. One infers that $\pi$ electrons are more delocalized than $\sigma$ electrons because only half as many orbitals cover the same available space.* It is also noteworthy that localized $\pi$ orbitals situated on joint atoms ($\pi\ell 2$, $\pi\ell 3$, $\pi\ell 4$, $\pi\ell 2'$) contribute more stabilization than those located on non-joint atoms, i.e. the joint provides more paths for local delocalization.

## 6.12. Localization Including Sigma Electrons

One can also localize aromatic systems by applying the localization procedure to $\pi$ electrons and $\sigma$ electrons simultaneously. The sigma electrons will then be localized in the regions of the single bonds. Since the localization energy to be gained from the $\pi$ electron localizations is less than that from the sigma electron localization, the total localization will be dominated by the latter. This leads to modifications of the localized $\pi$ orbitals. In benzene, for example, a Kekulé localization which mixes the $\sigma$ and $\pi$ orbitals to form double banana bonds is preferred over the other equivalent $\pi$ localizations discussed. [60] In naphthalene a Kekulé type structure is found similar to the one presently discussed, but different in that the $(\pi\ell 2)$ are hybridized with corresponding $\sigma$-CC bonding orbitals to form banana bonds, whereas the $(\pi\ell 2')$ remains a pure $\pi$ orbital. [61] While this is of interest in the discussion of the whole molecule, it is clear that certain intrinsic properties of the $\pi$-electrons are more readily recognized by the localization which has been discussed here. We hope to discuss elsewhere localized orbitals involving $\sigma$ bonds in organic molecules.

## Acknowledgment

The authors are particularly grateful to Kenneth Sundberg for plotting the *pi*-orbital contours.

## References

[1] Mulliken, R. S.: Phys. Rev. *32*, 186, 761 (1928); *41*, 49 (1932) and elsewhere. A discussion of the complete early work of Mulliken is given by Coulson, C. A.: In: Molecular Orbitals in Chemistry, Physics, and Biology, pp. 1–15. New York: Academic Press 1964.

[2] Pauling, L.: Proc. Natl. Acad. Sci. *14*, 359 (1928); J. Am. Chem. Soc. *53*, 1367, 3225 (1931) and elsewhere. A complete presentation is given in the book Pauling, L.: The Nature of the Chemical Bond. Ithaca, N.Y.: Cornell University Press 1960.

[3a] The invariance of determinantal wave functions was pointed out by Fock, V.: Z. Physik *61*, 126 (1930). Hund, F.: Z. Physik *73*, 1, 565 (1931 and 1932) was the first to formulate molecular orbital wavefunctions (for $H_2O$) using symmetry orbitals as well as localized orbitals; Coulson, C. A.: Trans. Faraday Soc. *33*, 388 (1937) used localized orbitals for the $CH_4$ molecule.

[3b] The first consistent discussion relating symmetry orbitals and localized orbitals through the determinantal invariance was given by Lennard-Jones, J. E.: Proc. Roy. Soc. (London) *A198*, 1, 14 (1949). This discussion concerned mainly equivalent orbitals in symmetric molecules. These were further investigated by Hall, G. G., and Lennard-Jones, J.E.: Proc. Roy. Soc. (London) *A202*, 155 (1950); Lennard-Jones, J. E., and Pople, J. A.: Proc. Roy. Soc. (London) *A202*, 166 (1950); and Hurley, A. C.: On Orbital Theories of Molecular Structure, Thesis, Trinity College, Cambridge, England 1952.

3c) Later contributions to the theory of localized orbitals were made by Löwdin, P.-O.: Phil. Mag. Suppl. *5*, 29 (1956); Boys, S. F.: Rev. Mod. Phys. *32*, 300 (1960); Ruedenberg, K.: Rev. Mod. Phys. *34*, 326 (1963), Sec. 3 and Peters, D.: J. Chem. Soc. 2003, 2015, 4017 (1963).

3d) The latest revival of the subject was by Edmiston, V., Ruedenberg, K.: Rev. Mod. Phys. *34*, 457 (1963) who developed a method for obtaining localized orbitals rigorously and investigated many systems [J. Chem. Phys. *43*, S97 (1965). Quantum Theory of Atoms, Molecules and Solid State, p. 263. New York, N.Y.: Academic Press 1966.

4) Fock, V.: Z. Physik *61*, 126 (1930).

5) Roothaan, C. C. J.: Rev. Mod. Phys. *23*, 69 (1951).

6) See Ref. 4.

7) See Ref. 5.

8) Ibid.

9) Ibid.

10) Ibid.

11) See, for example, Pilar, F. L.: Elementary Quantum Chemistry, McGraw-Hill, New York, N.Y., 1966, 236 ff.

12) See Ref. 5.

13) Koopmans, T.: Physica, *I*, 104 (1933). The physical reason for the agreement between orbital energies and ionization potentials was clarified by Mulliken, R. S.: J. Chim. Phys. *46*, 497 (1949).

14) Lennard-Jones, J. E., Pople, J. A.: Proc. Roy. Soc. (London), *A202*, 166 (1950). For a detailed discussion, see Edmiston, C., and Ruedenberg, K.: Rev. Mod. Phys. *35*, 457 (1963). Since then, a variety of other reasonable localization criteria have been proposed and investigated. A survey and discussion of this aspect of localization theory is given by Weinstein, H., Pauncz, R., Cohen, M.: Advances in Atomic and Molecular Physics, Vol. 7. New York: Academic Press 1971.

15) Ibid.

16) Edmiston, C., Ruedenberg, K.: Rev. Mod. Phys. *35*, 457 (1963); Ruedenberg, K.: Modern Quantum Chemistry, Vol. 1, p. 85. New York, N.Y.: Academic Press 1965.

17) Edmiston, C., Ruedenberg, K.: Rev. Mod. Phys. *35*, 457 (1963) and J. Chem. Phys. *43*, S97 (1965).

18) See Ref. 17.

19) See Ref. 17.

20) Thompson, H. B.: Inorg. Chem. *7*, 604 (1968). These ideas are further pursued in a recent paper by Allen, L. C.: Theoret. Chim. Acta (in press).

21) Gillespie, R. J.: J. Chem. Educ. *40*, 295 (1963); J. Am. Chem. Soc. *82*, 5978 (1960); J. Chem. Soc. 4672 (1963).

22) Walsh, A. D.: J. Chem. Soc. 2260, 2266, 2288, 2296, 2301, 2306 (1953).

23) An example is the case of pyrene, to be discussed in Section 6 below.

24) Harriman, J. E. and Del Bene, J.: Univ. Wisconsin, Theor. Chem. Institute, Techn. Report No 340, 20 March 1969.

25) Edmiston, C., Ruedenberg, K.: Quantum Theory of Atoms, Molecules, and Solid State, p. 263. New York, N.Y.: Academic Press 1966.

26) See Ref. 17.

27) See Ref. 25.

28) Hall, G. G., Lennard-Jones, J. E.: Proc. Roy. Soc. (London) *A202*, 155 (1950).

29) Lennard-Jones, J. E.: Proc. Roy. Soc. (London) *A198*, 1, 14, (1949); Hall, G. G.: Proc. Roy. Soc. (London) *A202*, 336 (1950).

30) Ibid.

31) Lennard-Jones, J. E.: Proc. Roy. Soc. (London) *A198*, 1, 14 (1949); Coulson, C. A.: Trans. Faraday Soc. *33*. 388 (1937).

32) Lennard-Jones, J. E.: Proc. Roy. Soc. (London) *A198*, 1, 14 (1949); Pople, J. A.: Proc. Roy. Soc. (London) *A202*, 323 (1950).

33) An example is the case of $C_2$ investigated in reference 25.

34) The canonical orbitals used are optimal "minimal-basis-set molecular orbitals", as given by Ransil, B. J.: Rev. Mod. Phys. *32*, 239 (1960). The contours of exact SCF-AO's are not significantly different.

35) Edmiston, C., Ruedenberg, K.: J. Chem. Phys. *43*, S 97 (1965). These localized orbitals are based on the canonical orbitals given in Ref. 34. Whenever possible, the BMO orbitals are plotted; otherwise the contours are those of the SAO orbitals.

36) See Ref. 35.

37) Ibid.

38) Ibid.

39) See Ref. 21.

40) See Ref. 35.

41) Ibid.

42) Ibid.

43) Ibid.

44) Herzberg, G.: Molecular Spectra and Molecular Structure, p. 556. New York, N.Y.: D. Van Nostrand 1950.

45) See Ref. 17.

46) Bloch, F.: Z. Physik *52*, 555 (1928).

47) Coulson, C. A., Longuet-Higgins, H. C.: Proc. Roy. Soc. (London) *A191*, 39 (1947); *A192*, 16 (1947); *A193*, 447, 456 (1948); *A195*, 188 (1948); Mulliken, R. S., Rieke, C., Brown, S.: J. Am. Chem. Soc. *63*, 41 (1941); Mulliken, R.S., Rieke, C.: J. Am. Chem. Soc. *63*, 1770 (1941); Wheland, G. W.: J. Am. Chem. Soc. *63*, 2025 (1941). This method is described as the tight-binding approximation by Ruedenberg, K.: J. Chem. Phys. *34*, 1861 (1961).

48) Ruedenberg, K., Scherr, C. W.: J. Chem. Phys. *21*, 1565 (1953); Ruedenberg, K.: J. Chem. Phys. *22*, 1878 (1954); Ham, N. S., Ruedenberg, K.: J. Chem. Phys. *29*, 1199 (1958). See also in: Free-Electron Theory of Conjugated Molecules. New York, N.Y.: John Wiley and Sons 1964.

49) Wannier, G.: Phys. Rev. *52*, 191 (1937).

50) See Ref. 47.

51) Ibid.

52) Ibid.

53) Mulliken, R. S.: J. Chim. Phys. *46*, 500, 521 (1949). See also Ruedenberg, K.: J. Chem. Phys. *19*, 1433 (1951).

54) England, W., Ruedenberg, K.: Theoret. Chim.Acta *22*, 196 (1971).

55) Kekulé, A.: Ber. *23*, 1265 (1890); *29*, 1971 (1896); J. Chem. Soc. *73*, 97 (1898).

56) See Ref. 25. See also: England, W.: Int. Journ. Quantum Chemistry (in press) for a general discussion of such "localization degeneray."

57) Ibid.

58) See the second paper in Ref. 16.

59) See for example, Fieser, L., Fieser, M.: Advanced Organic Chemistry, p. 880. New York, N.Y.: Rheinhold 1961.

60) England, W., Gordon, M.: J. Am. Chem. Soc. *91*, 6864 (1969).

61) Unpublished calculation by England, W., and Gordon, M.

Received December 3, 1970

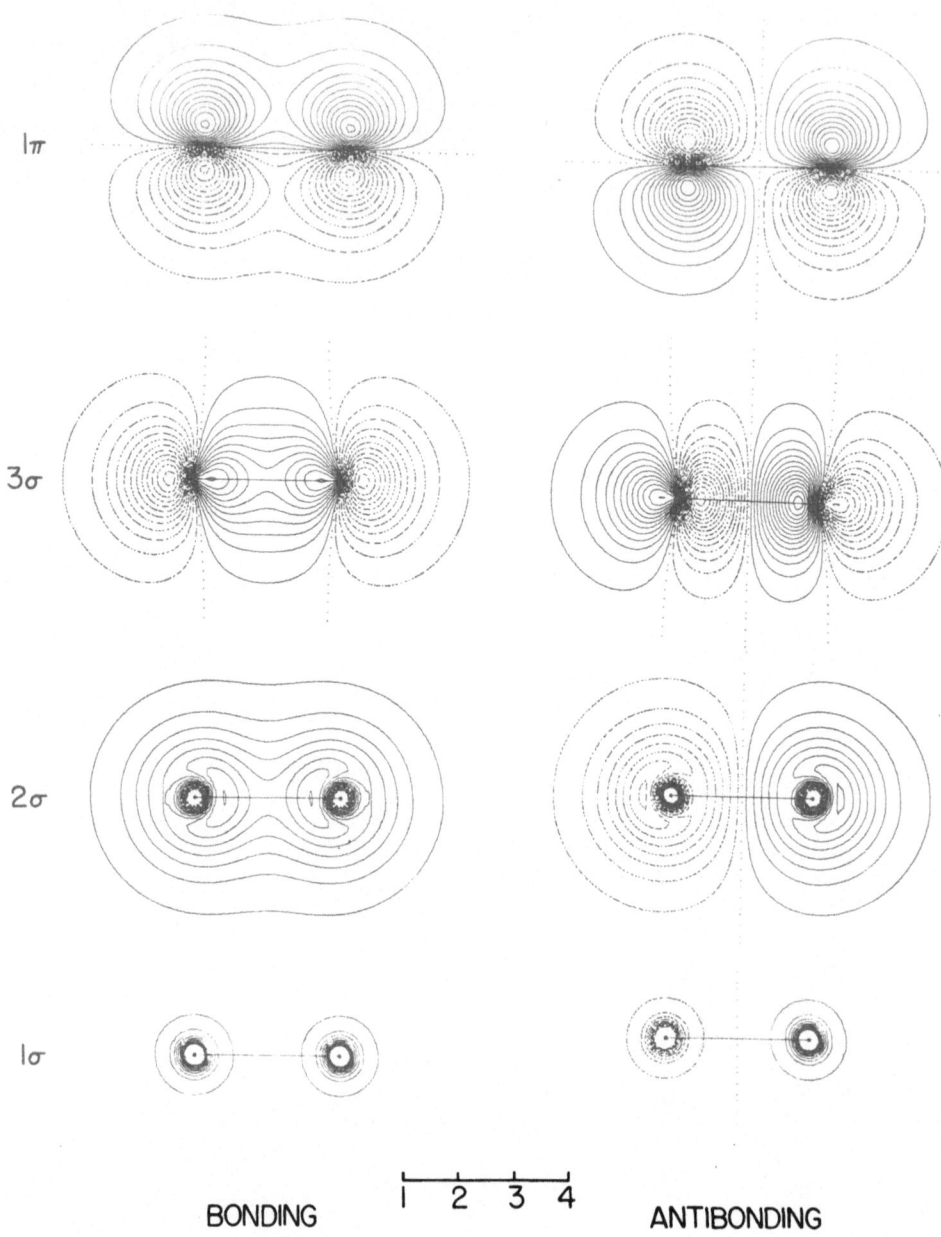

Fig. 1. Occupied canonical MO's in homonuclear diatomic molecules. Scale in this and subsequent figures in Bohr radii.

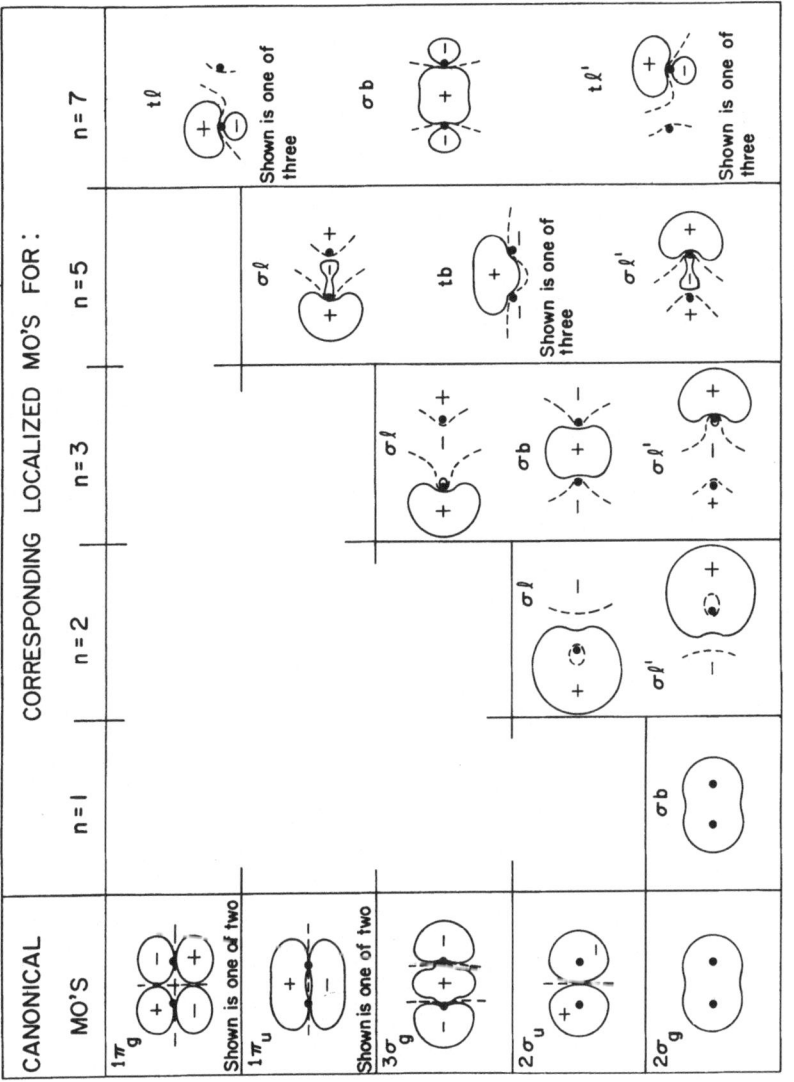

Fig. 2. Schematic diagrams of occupied canonical MO's and the corresponding localized MO's in homonuclear diatomic molecules. The number of doubly occupied valence orbitals is denoted by $n$

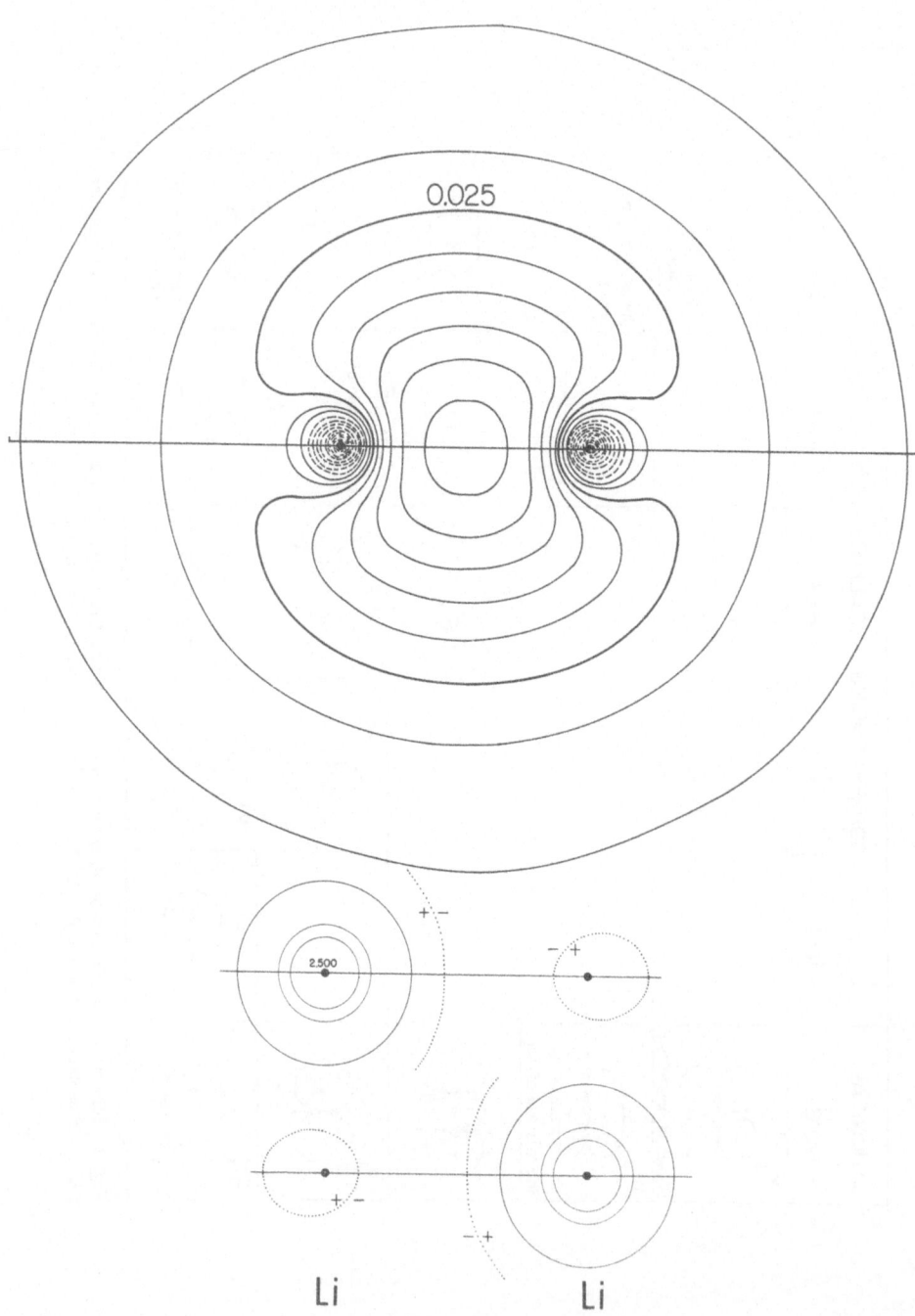

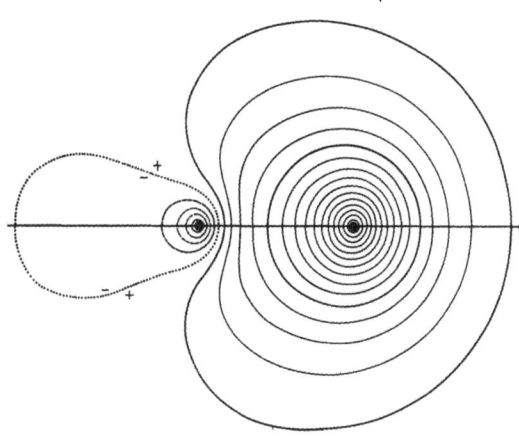

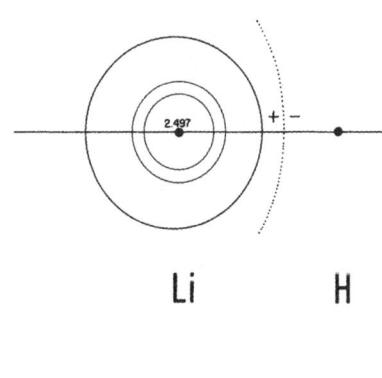

Li          H

0   1          5

Fig. 3.  Localized MO's
in Li$_2$ and LiH.

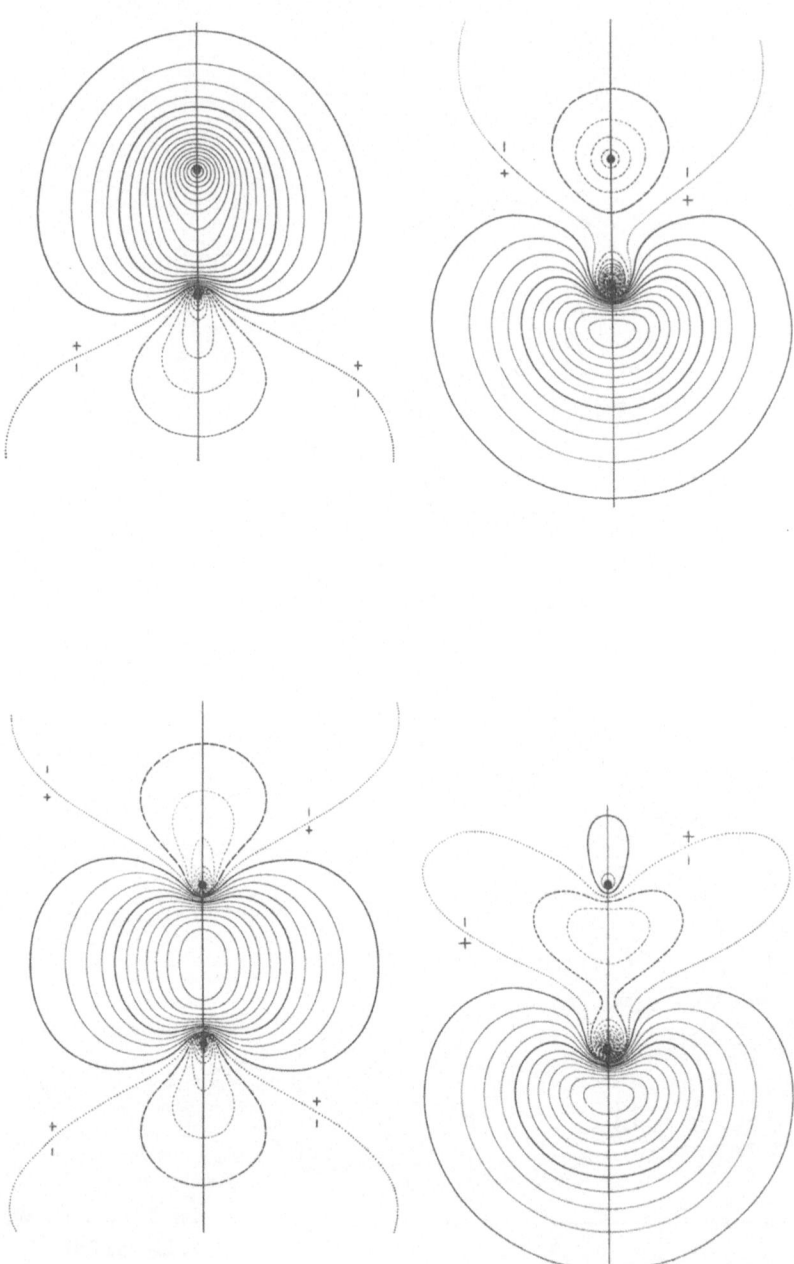

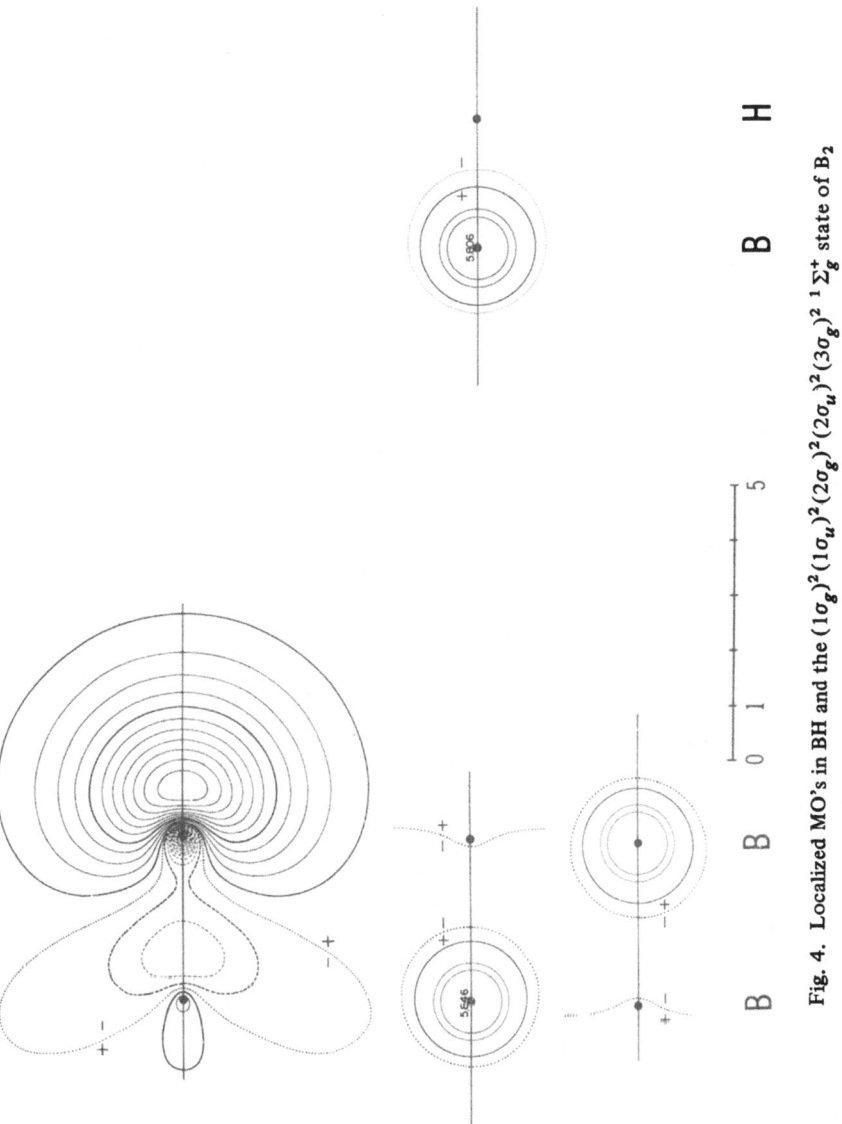

Fig. 4. Localized MO's in BH and the $(1\sigma_g)^2(1\sigma_u)^2(2\sigma_g)^2(2\sigma_u)^2(3\sigma_g)^2\ {}^1\Sigma_g^+$ state of $B_2$

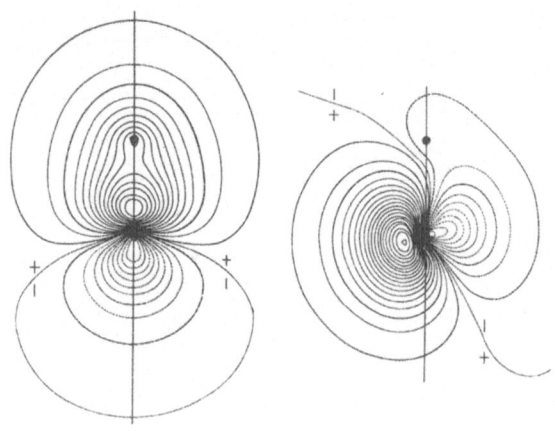

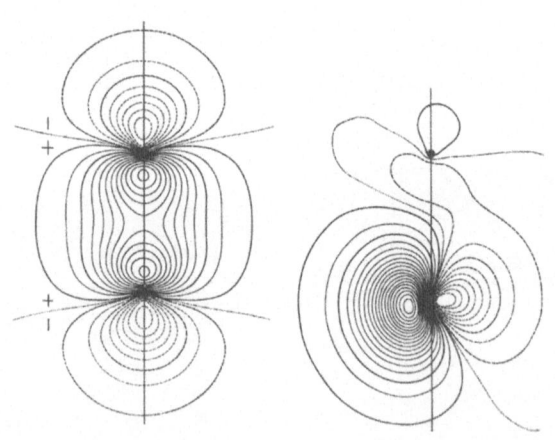

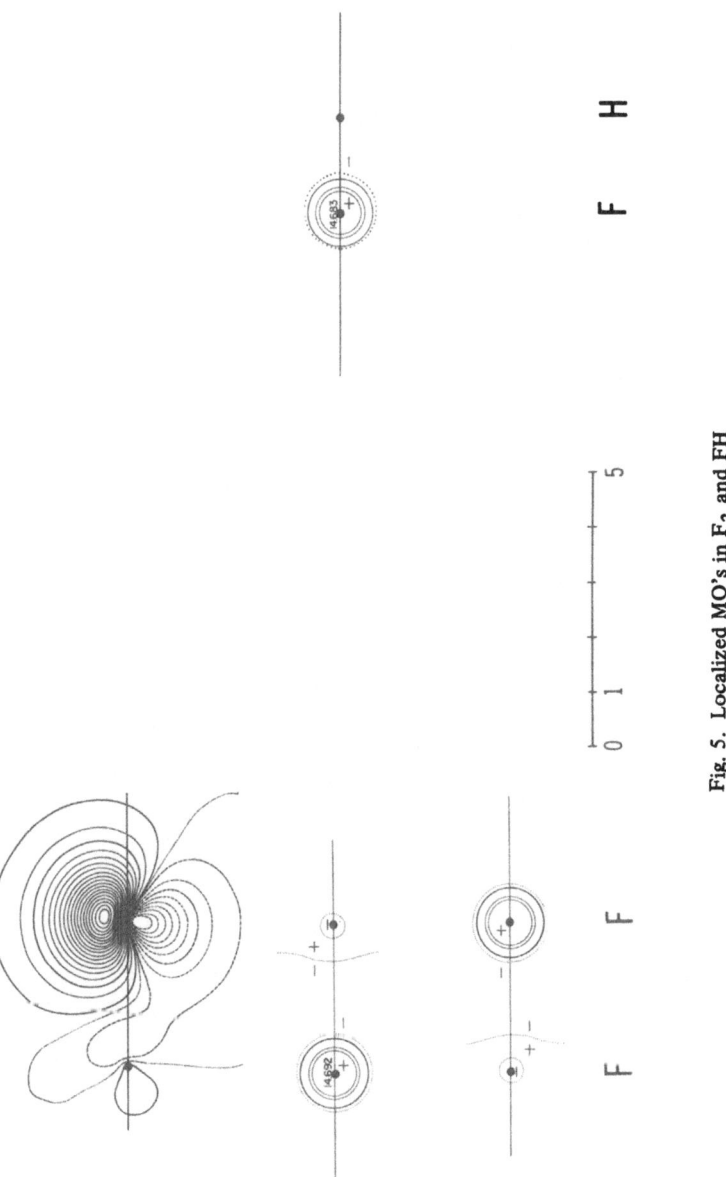

Fig. 5. Localized MO's in $F_2$ and FH

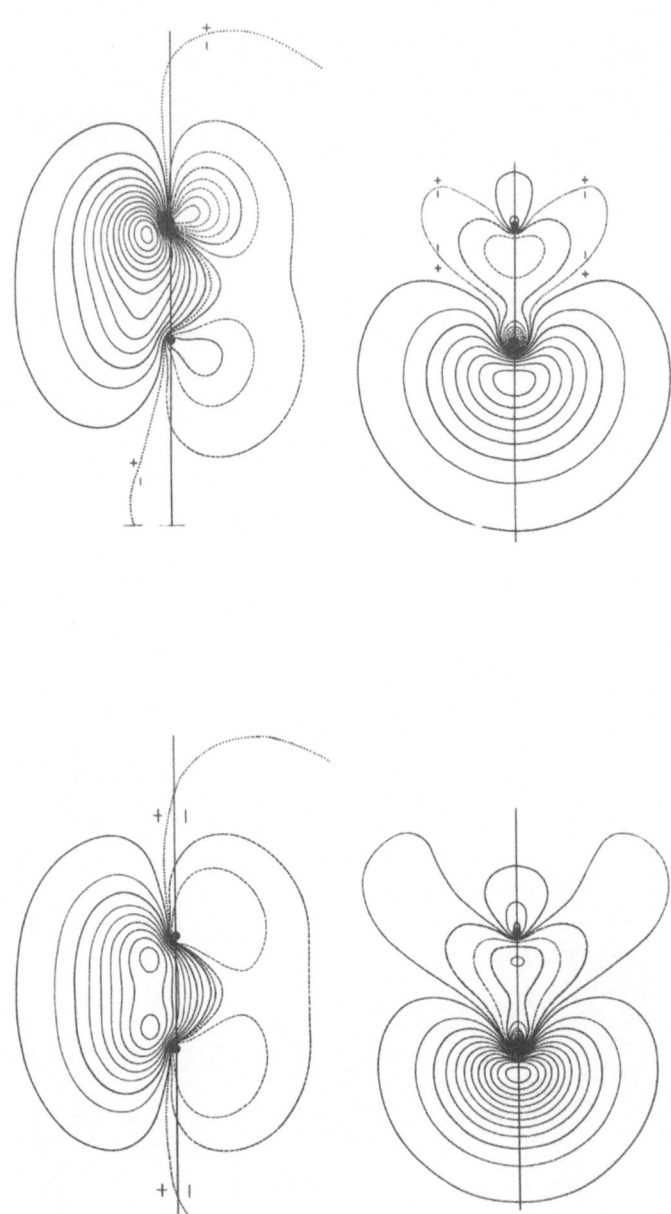

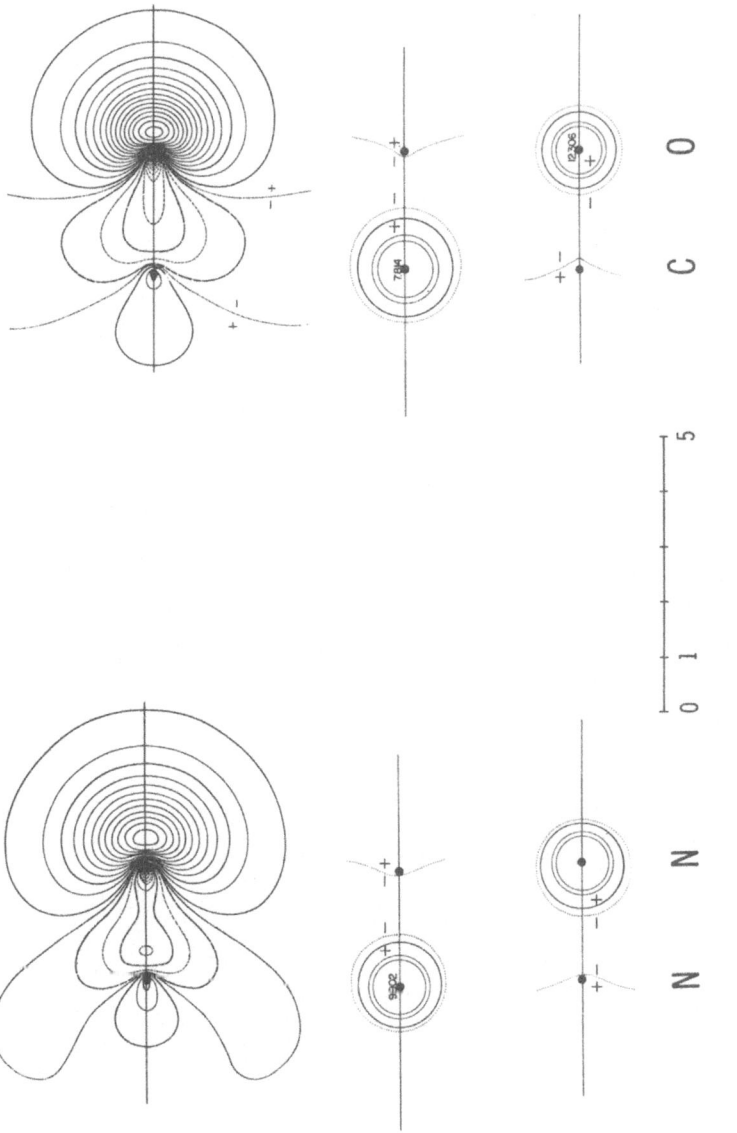

Fig. 6. Localized MO's in CO and $N_2$

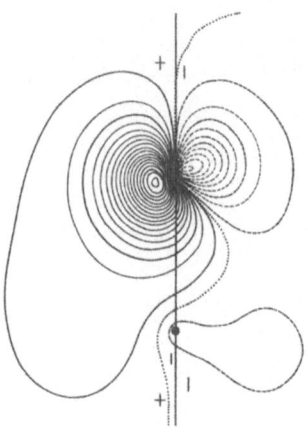

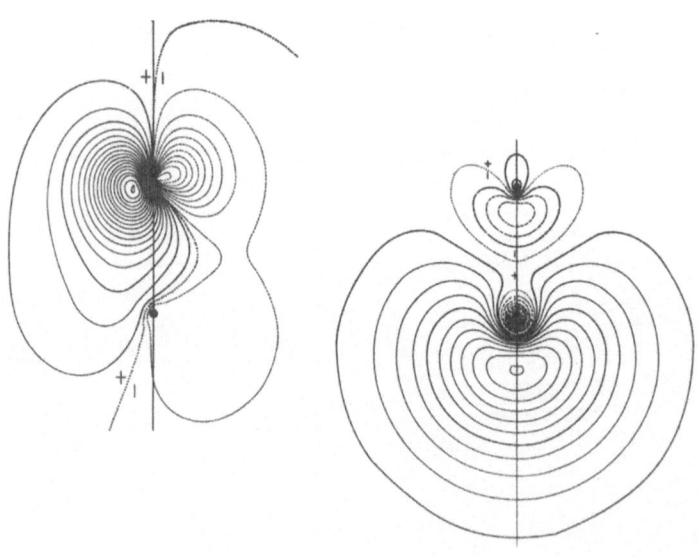

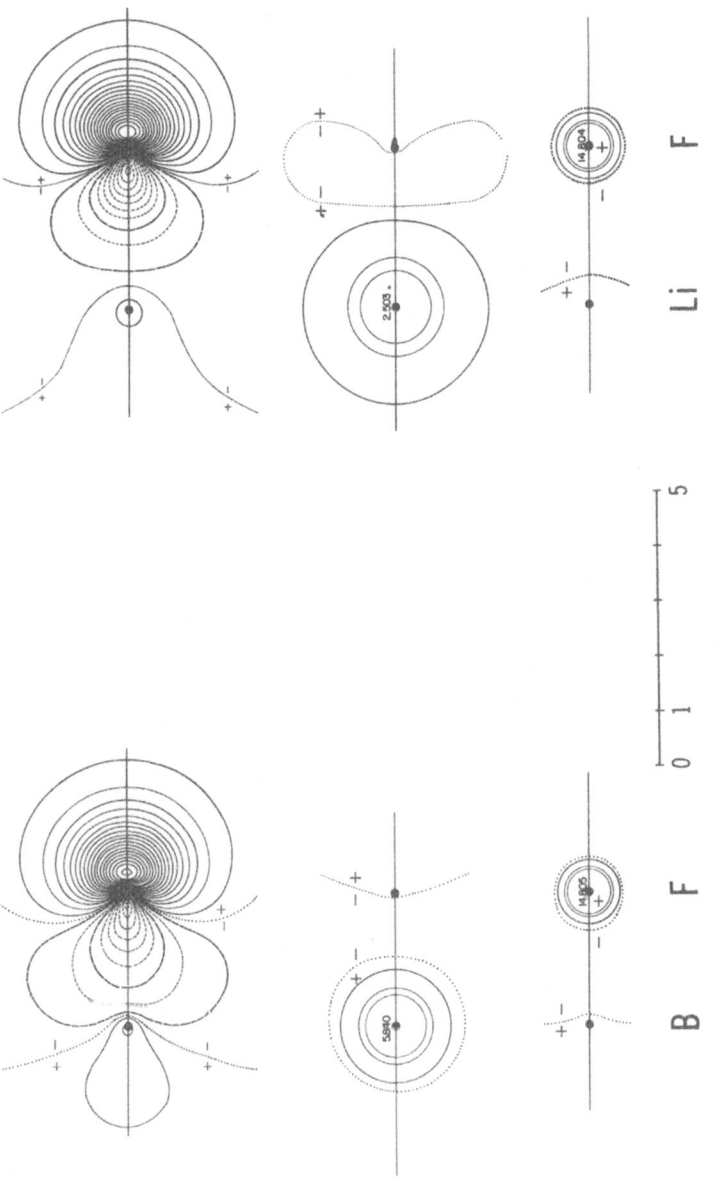

Fig. 7. Localized MO's in BF and LiF

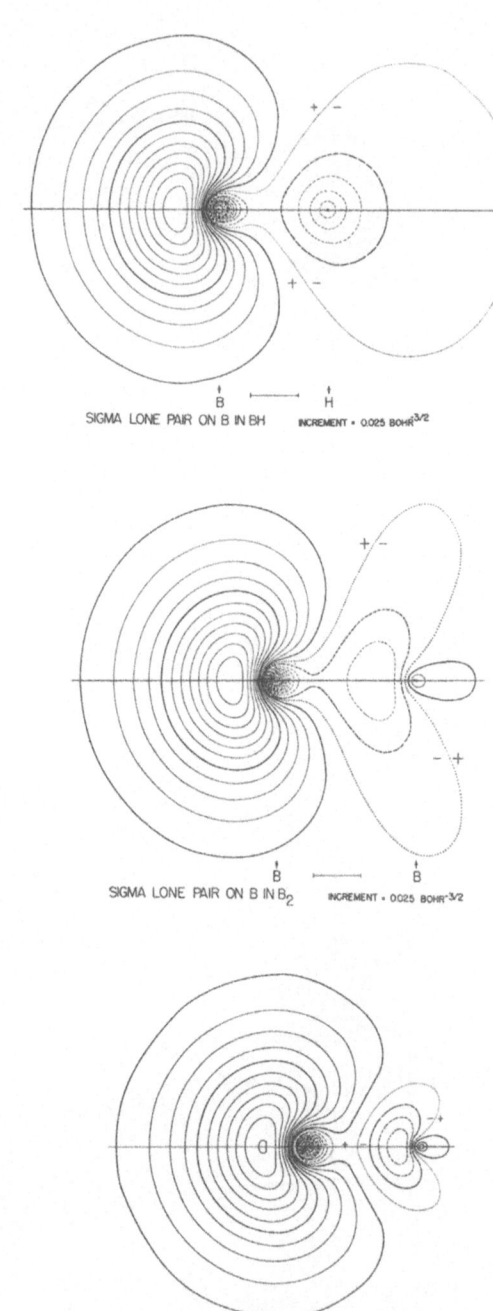

SIGMA LONE PAIR ON B IN BH    INCREMENT = 0.025 BOHR$^{-3/2}$

SIGMA LONE PAIR ON B IN B$_2$    INCREMENT = 0.025 BOHR$^{-3/2}$

SIGMA LONE PAIR ON B IN BF    INCREMENT = .025 BOHR$^{-3/2}$

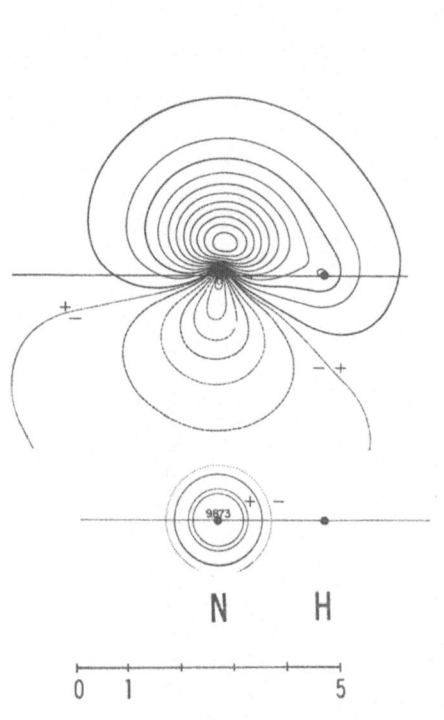

N    H

0    1                    5

Fig. 8. Localized MO's in the
$^1\Sigma (1\sigma)^2 (2\sigma)^2 (1\pi)^4$ excited
state of NH

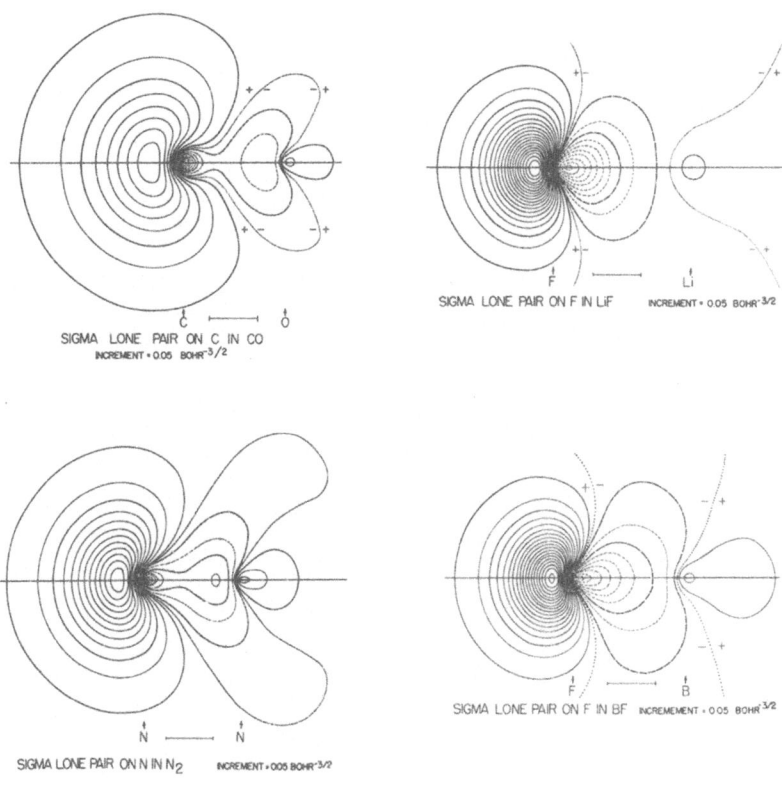

SIGMA LONE PAIR ON C IN CO
INCREMENT = 0.05 BOHR$^{-3/2}$

SIGMA LONE PAIR ON F IN LiF    INCREMENT = 0.05 BOHR$^{-3/2}$

SIGMA LONE PAIR ON N IN N$_2$    INCREMENT = 0.05 BOHR$^{-3/2}$

SIGMA LONE PAIR ON F IN BF    INCREMENT = 0.05 BOHR$^{-3/2}$

SIGMA LONE PAIR ON O IN CO
INCREMENT = 0.05 BOHR$^{-3/2}$

Fig. 9. Sigma lone pair MO's for diatomic molecules.

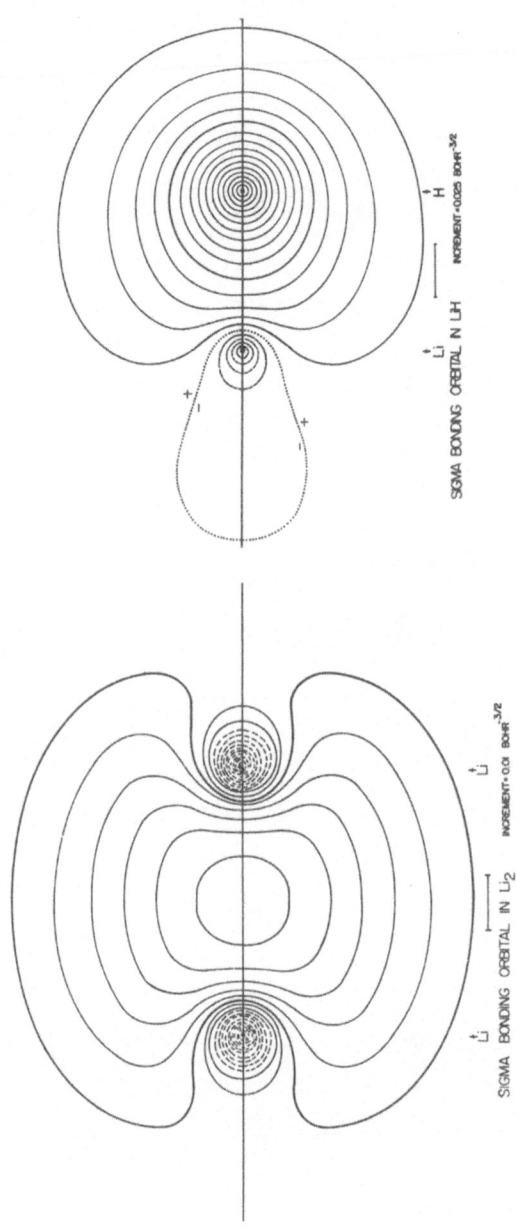

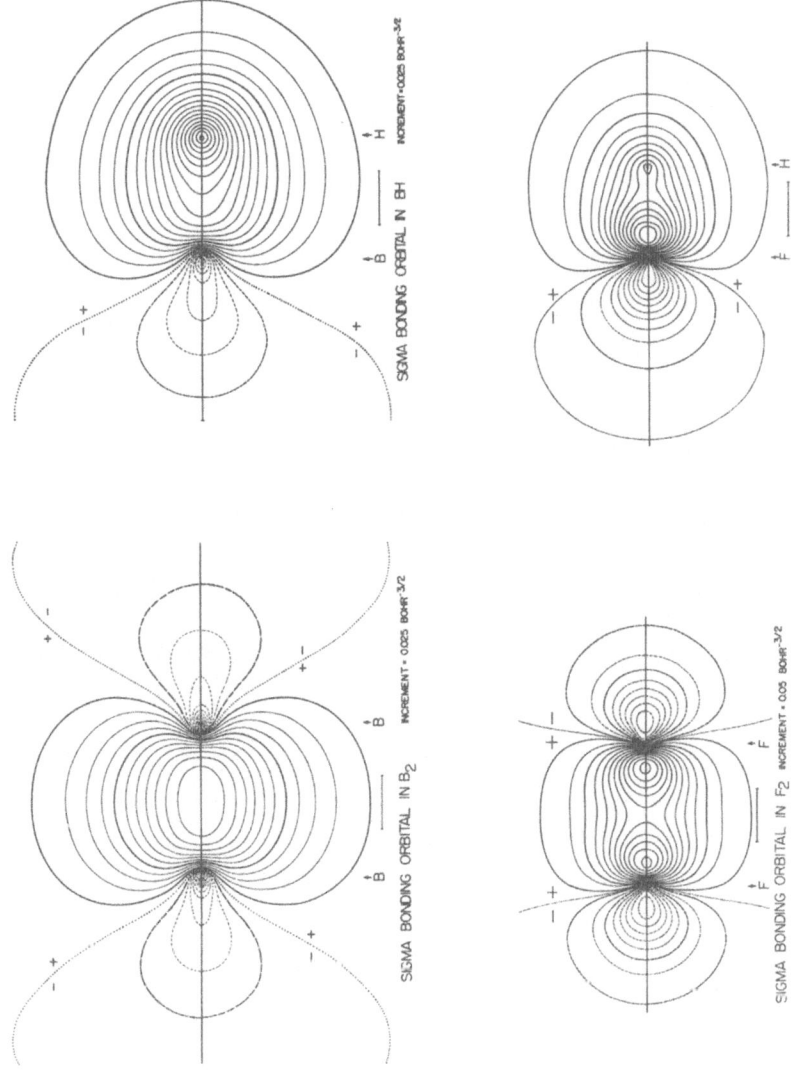

Fig. 10. Sigma bonding MO's for diatomic molecules

85

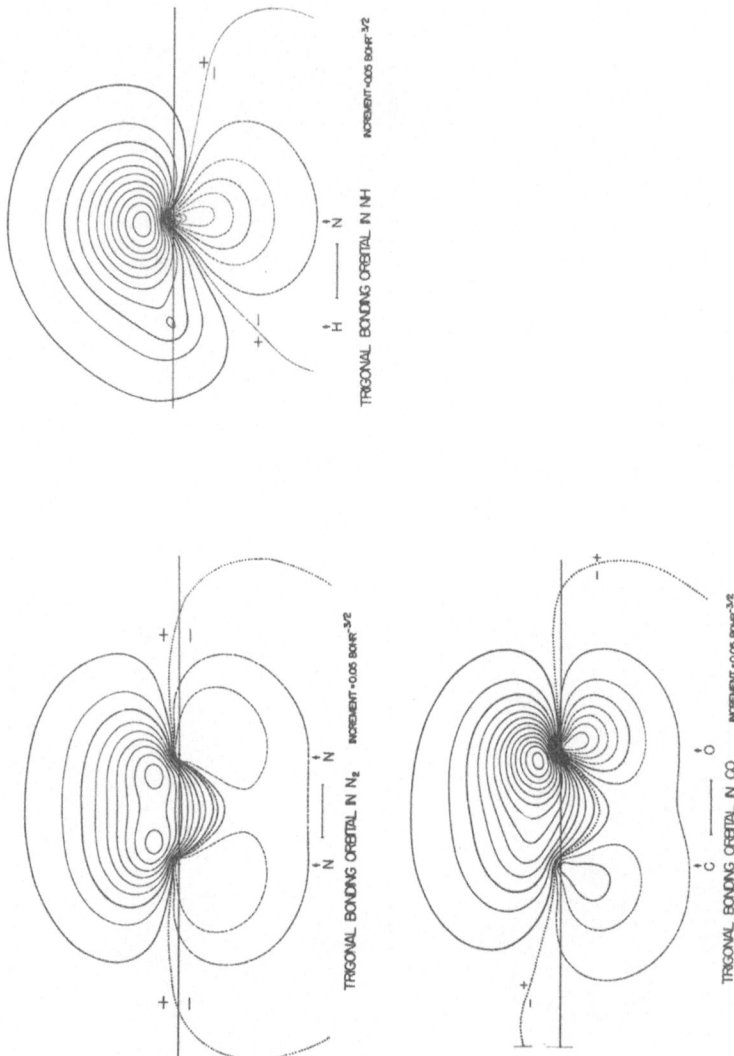

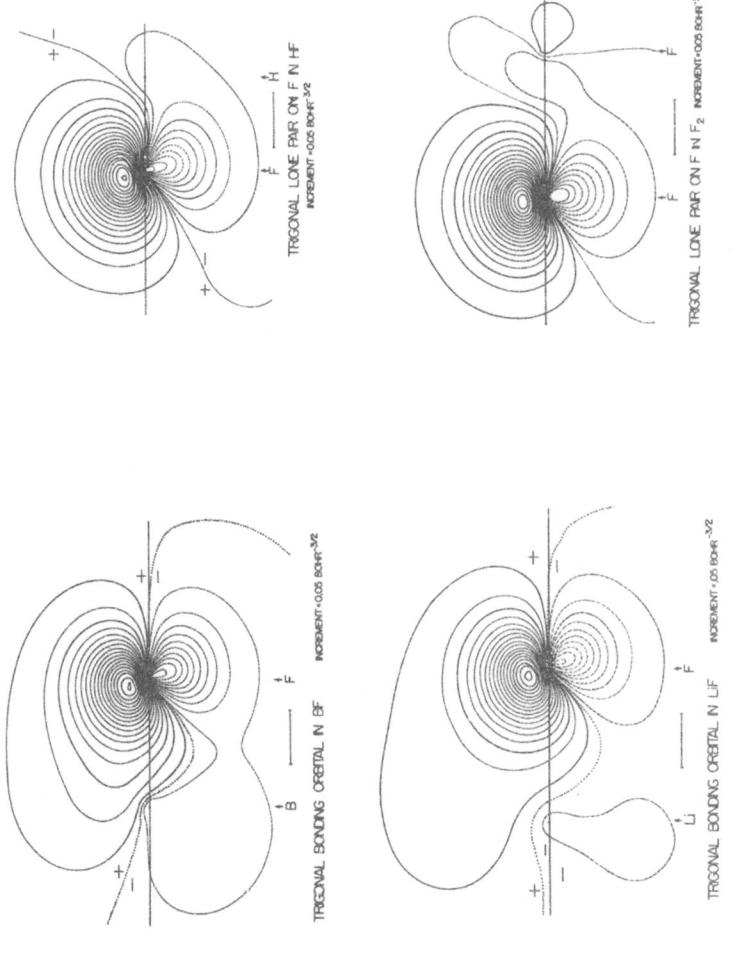

Fig. 11. Trigonal bonding and lone pair MO's for diatomic molecules.

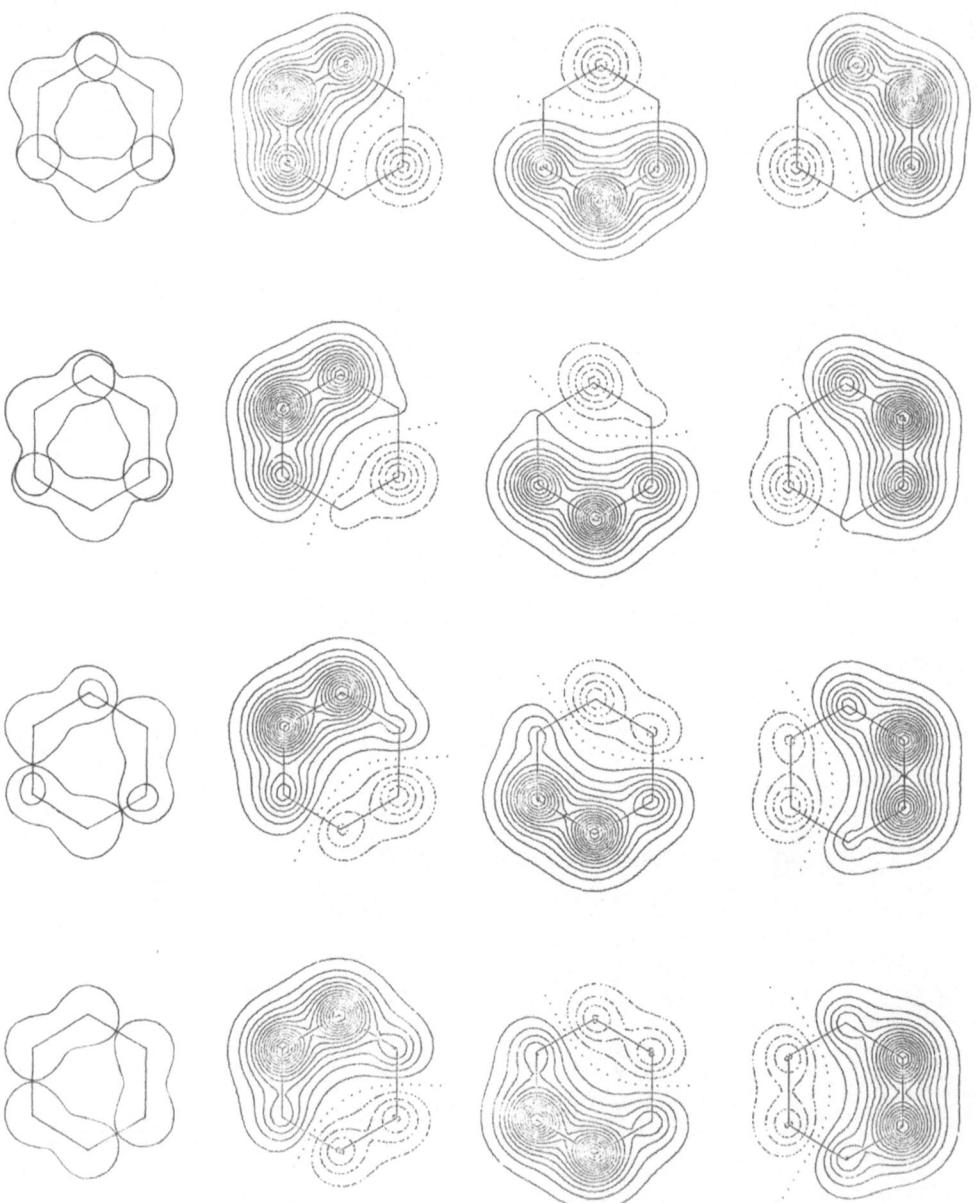

Fig. 12. Localized MO's for the benzene $\pi$-system

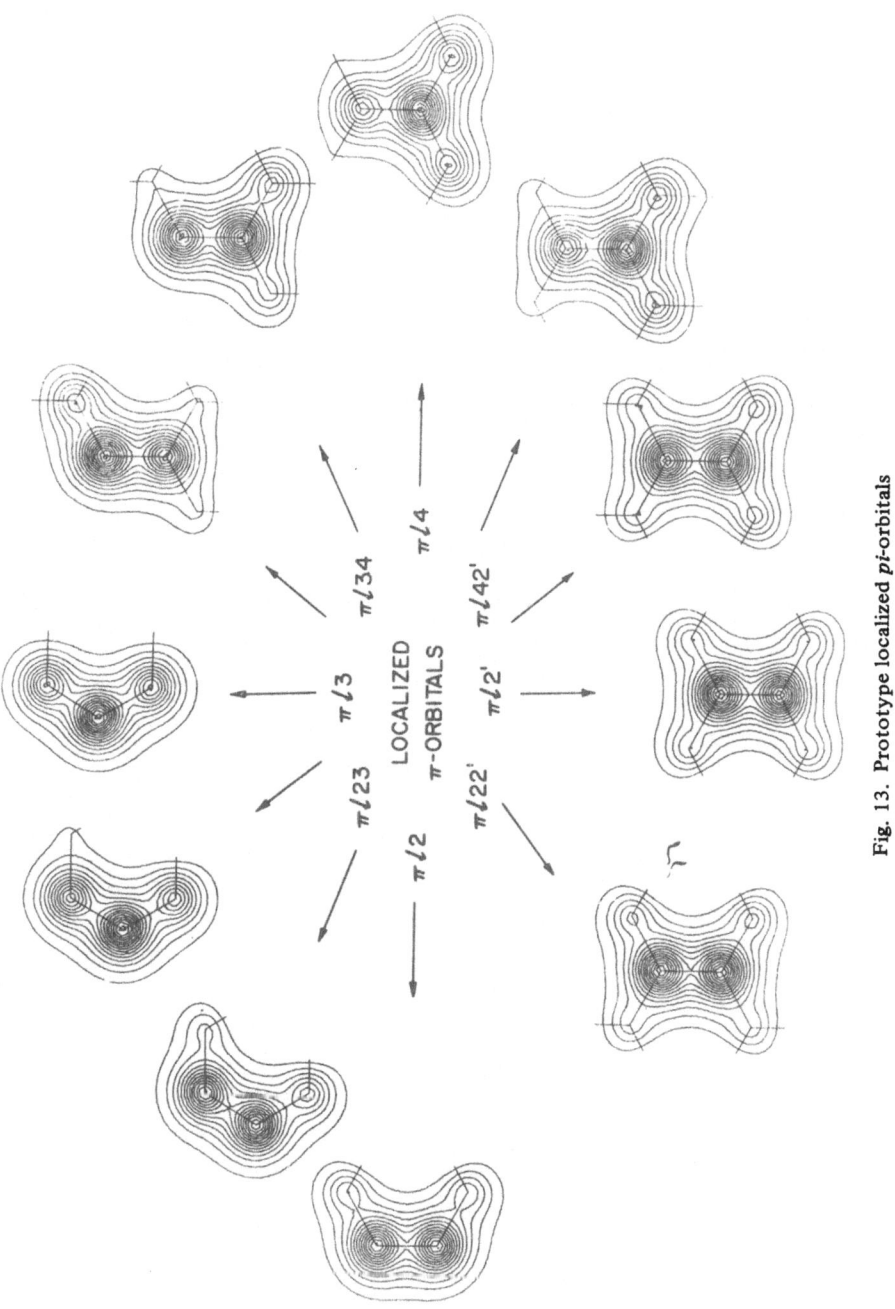

Fig. 13. Prototype localized *pi*-orbitals

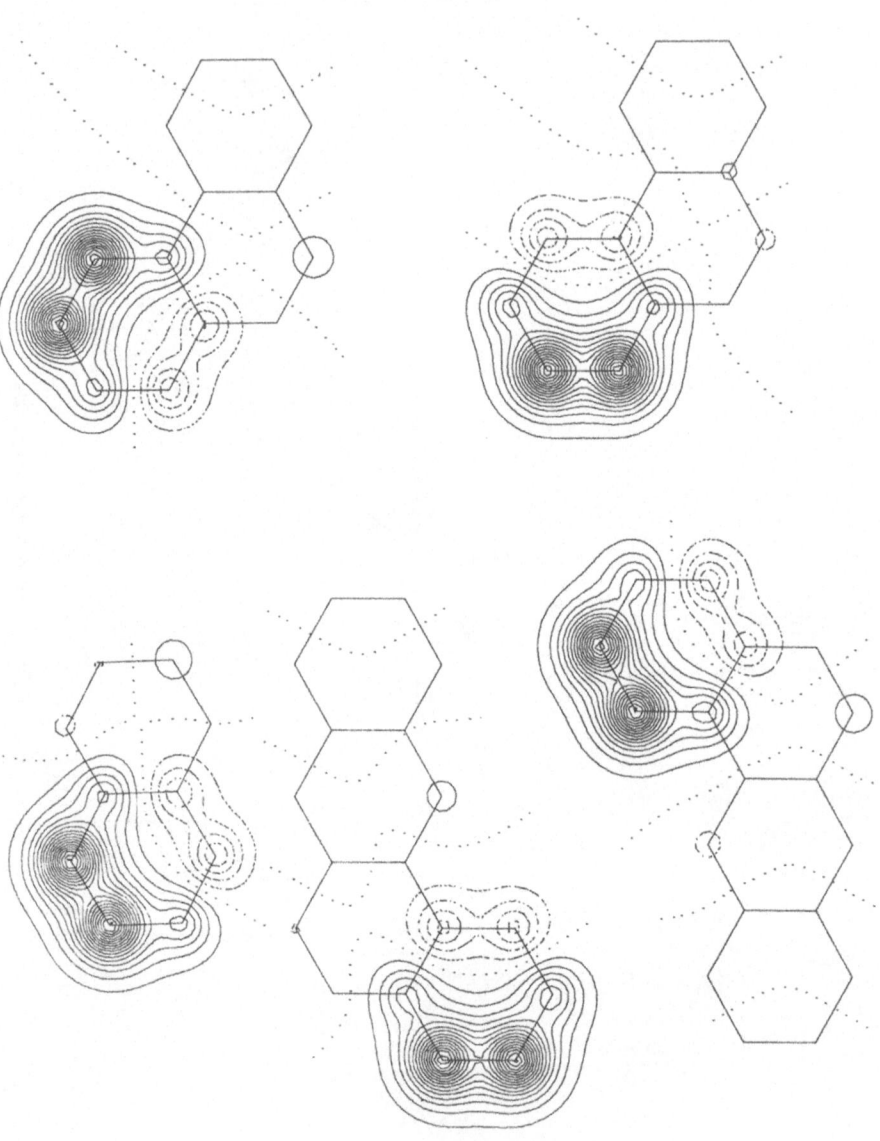

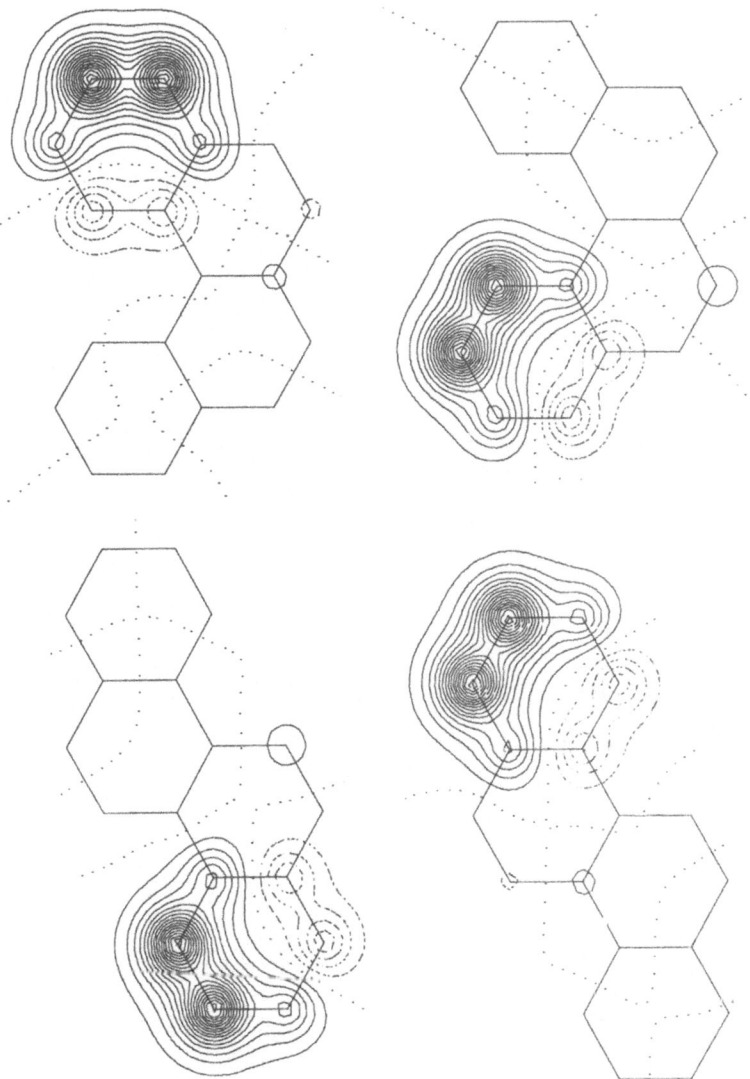

Fig. 14. $\pi\ell 2$ orbitals on branches of four non-joint atoms

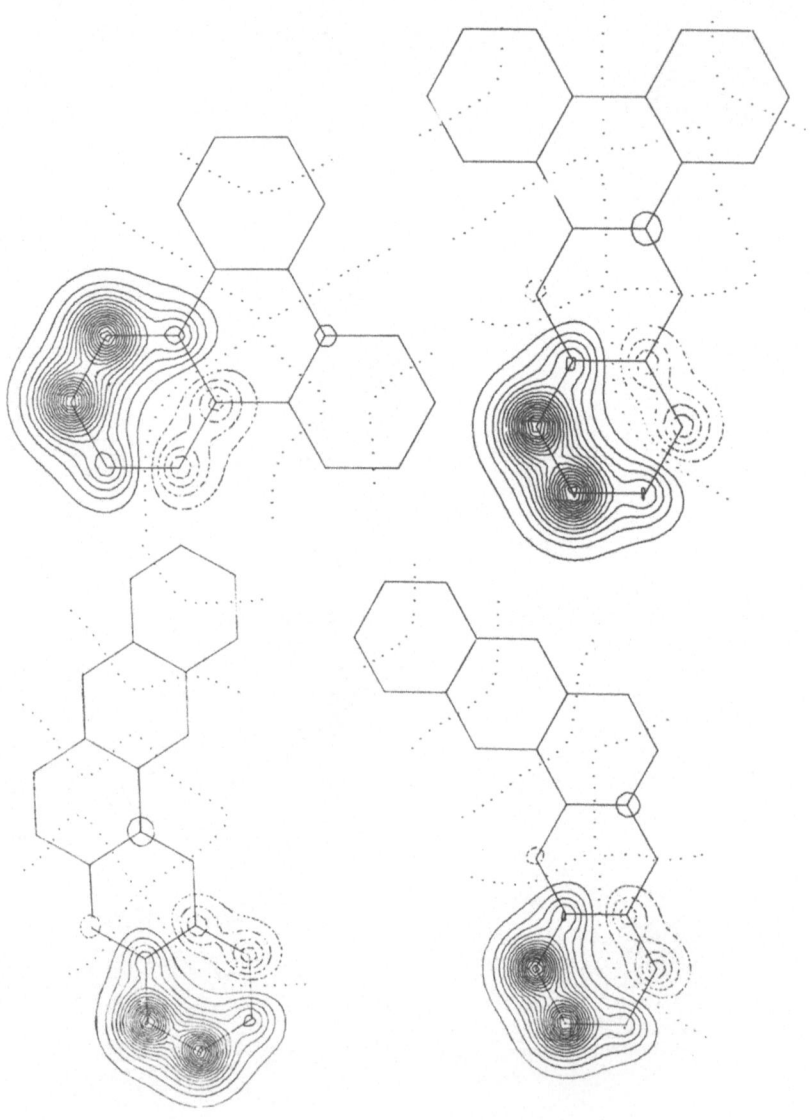

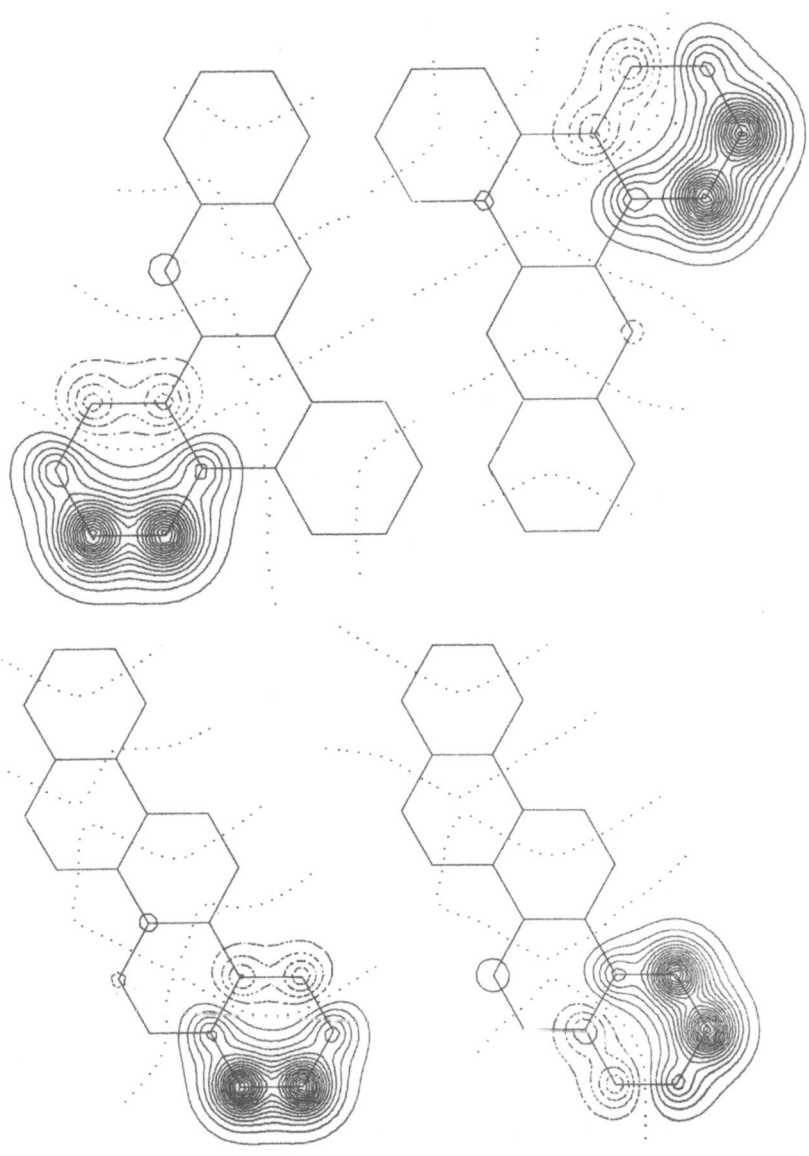

Fig. 15. $\pi\ell 2$ orbitals on branches of four non-joint atoms.

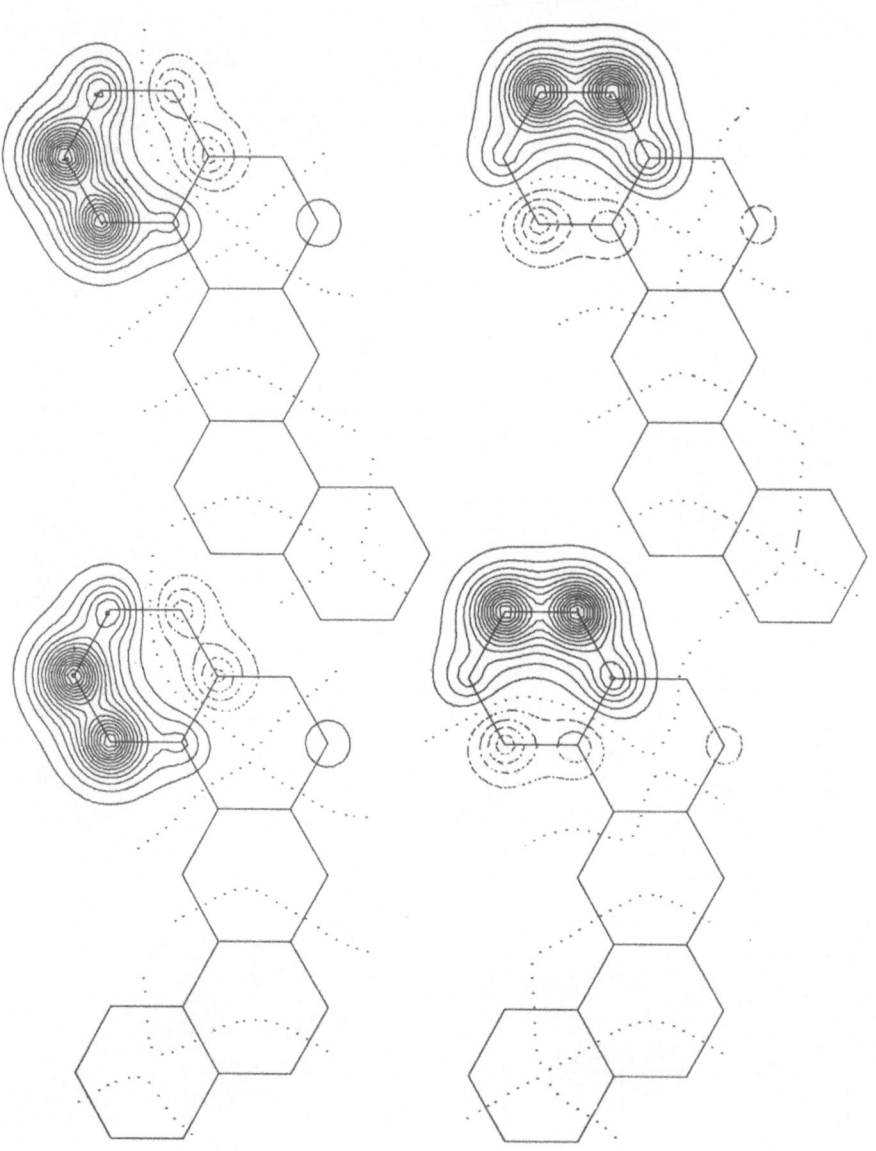

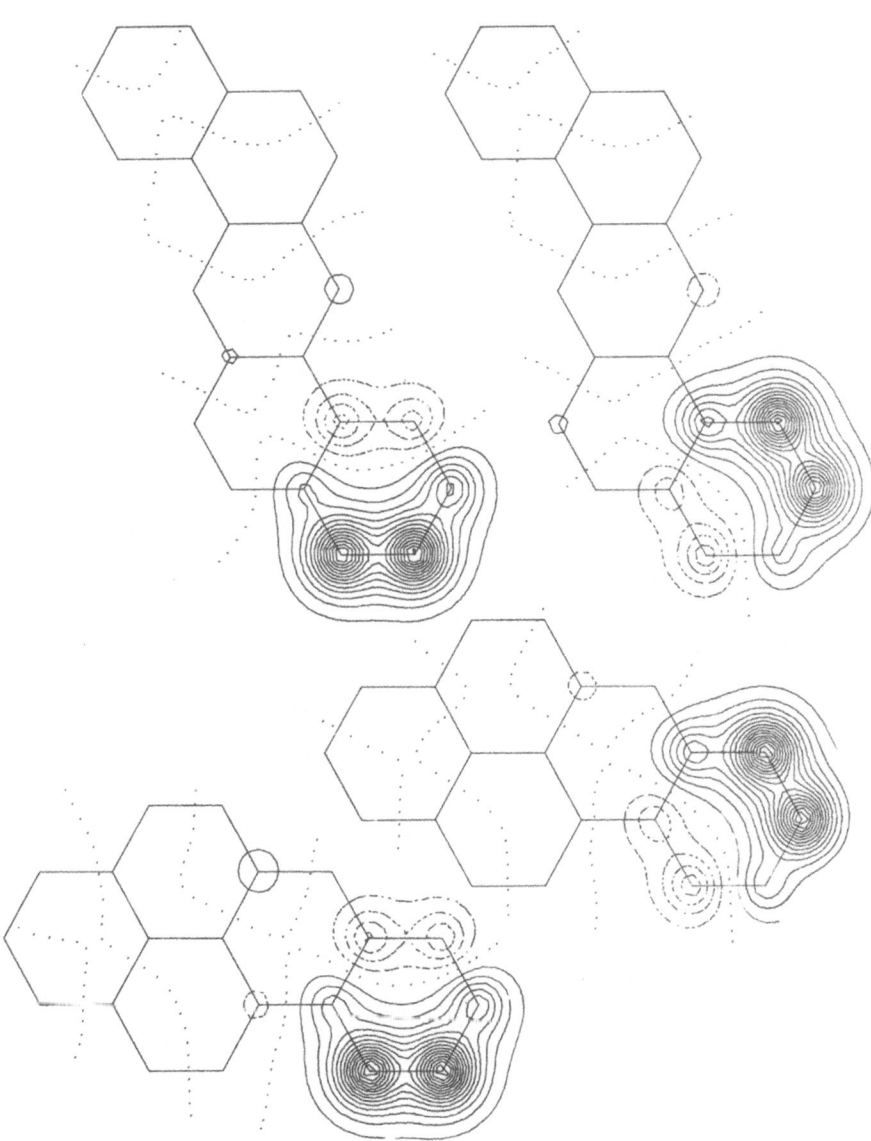

Fig. 16. $\pi\ell 23$ orbitals on branches of four non-joint atoms.

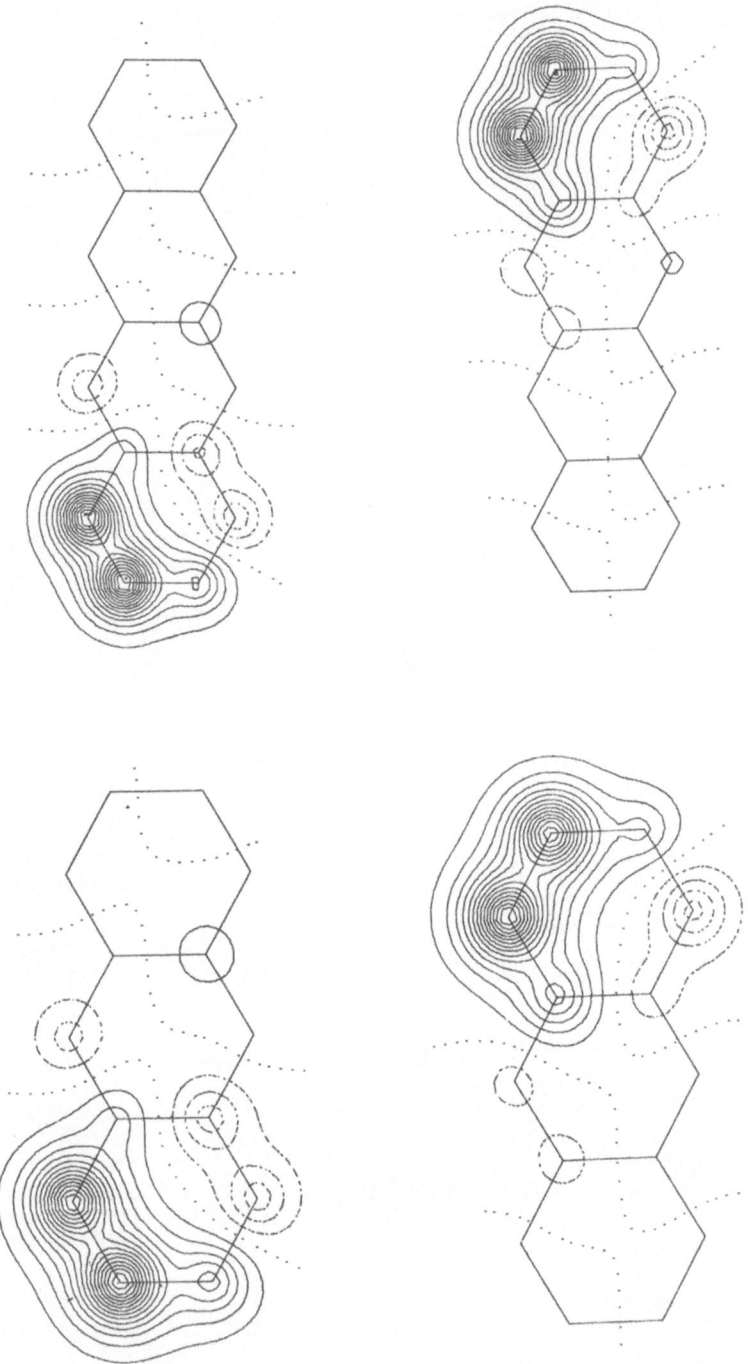

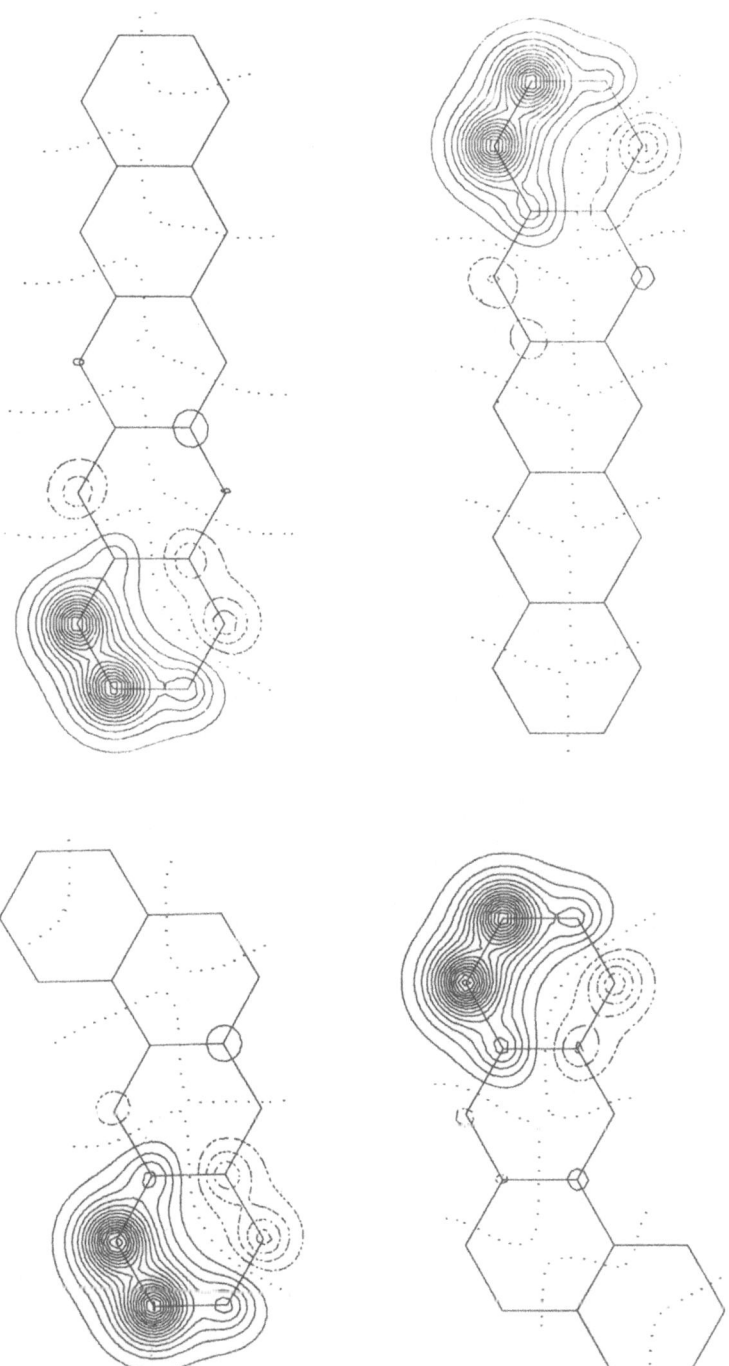

Fig. 17. $\pi\ell23$ orbitals on branches of four non-joint atoms

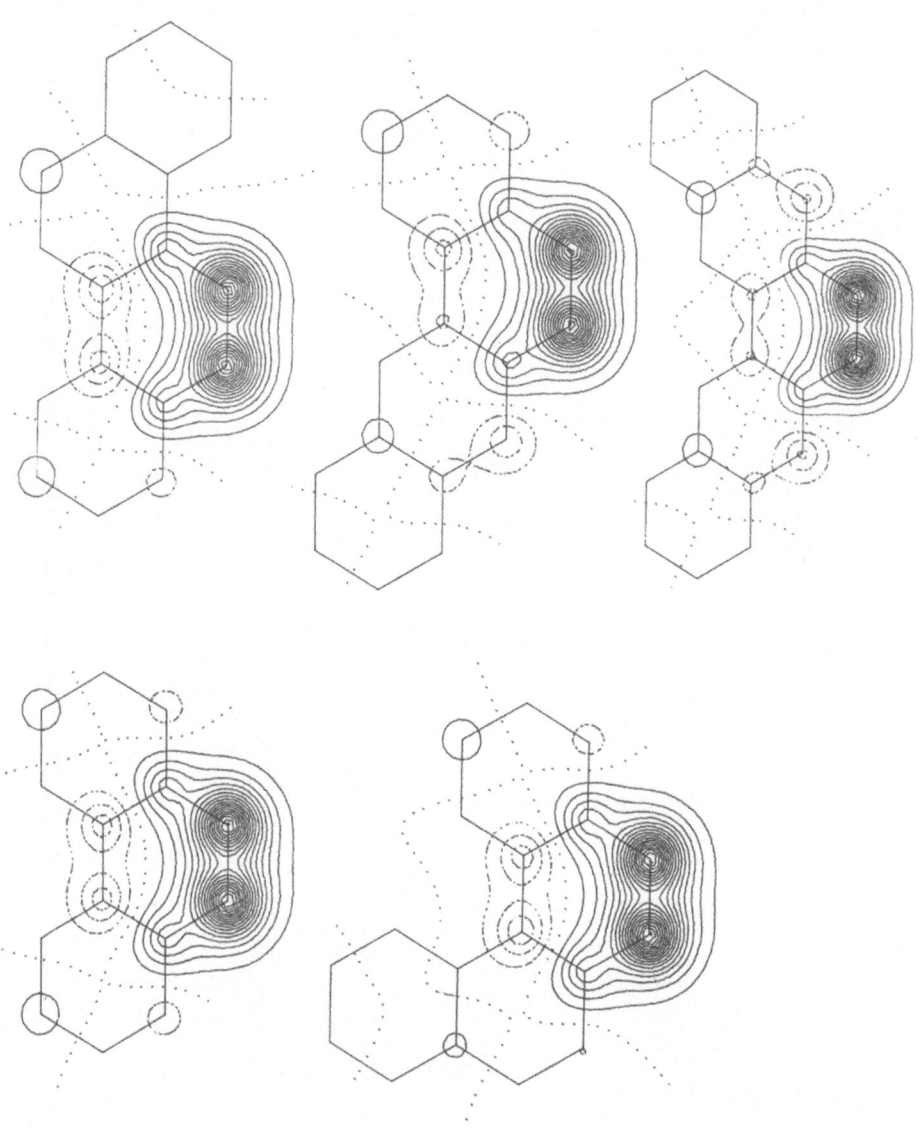

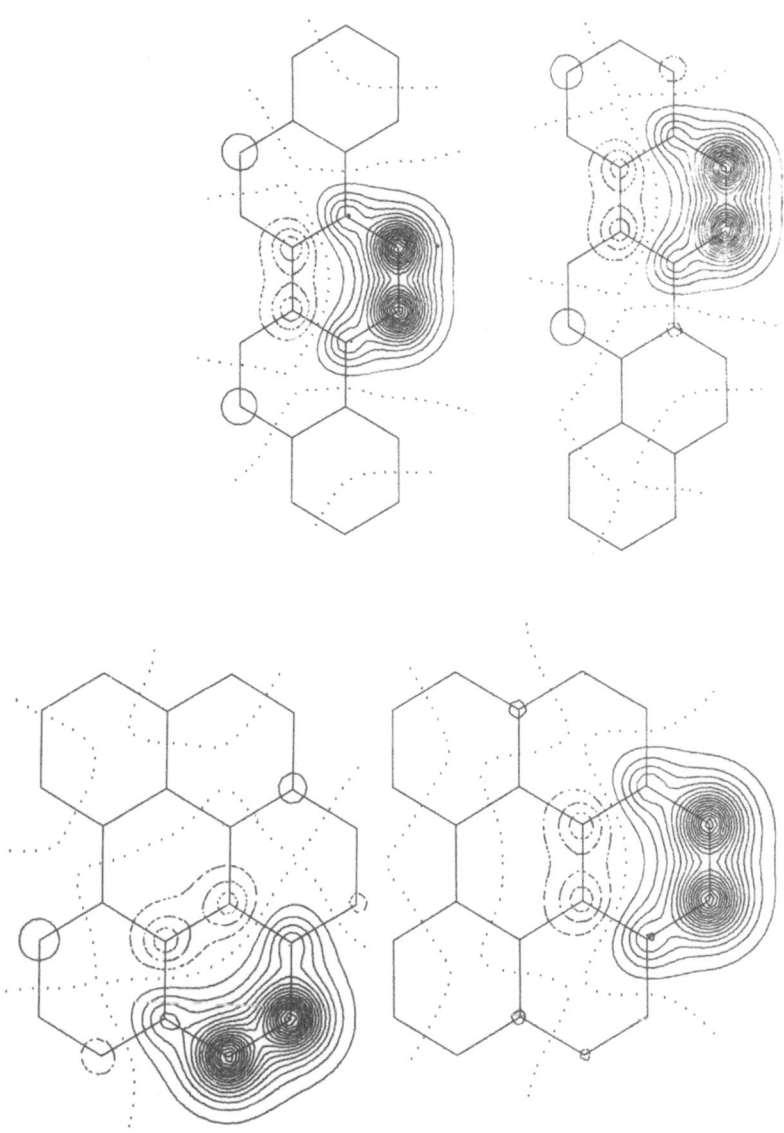

Fig. 18. $\pi\ell 2$ orbitals on branches of two non-joint atoms.

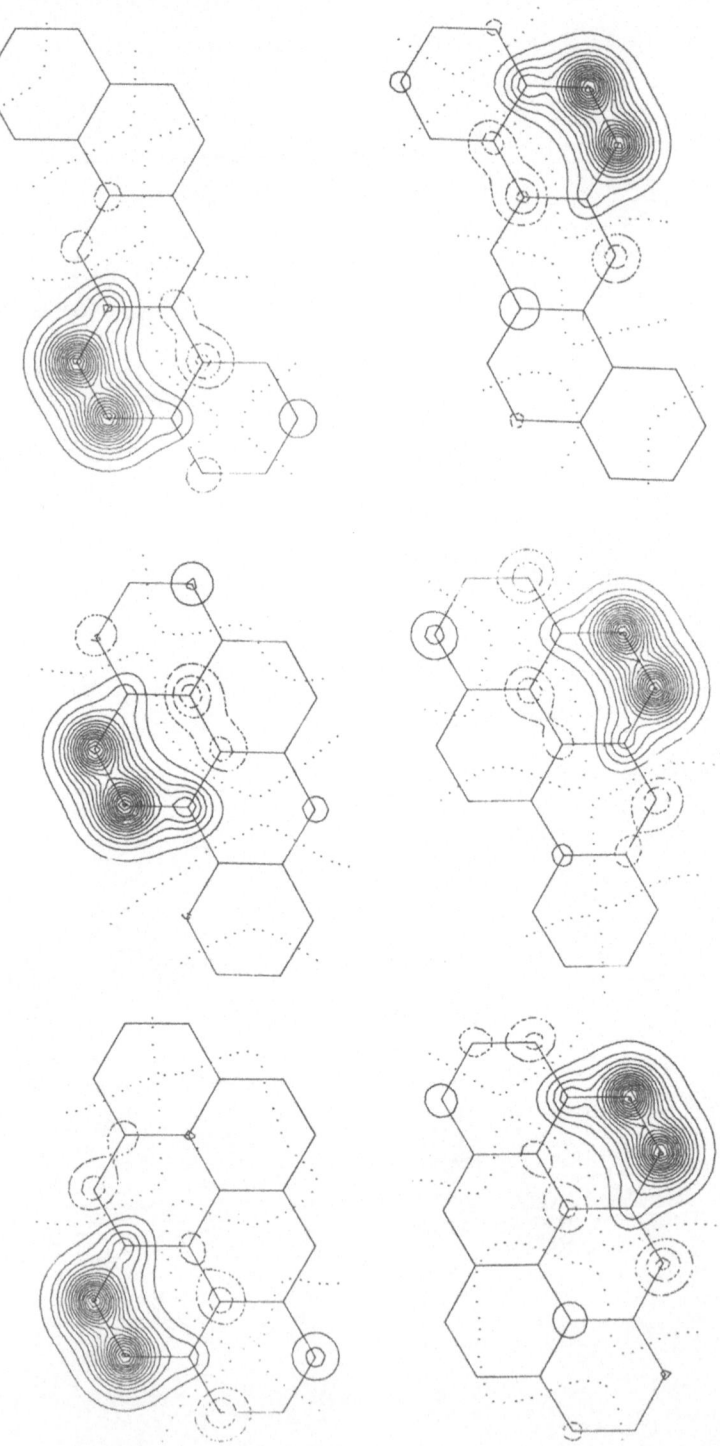

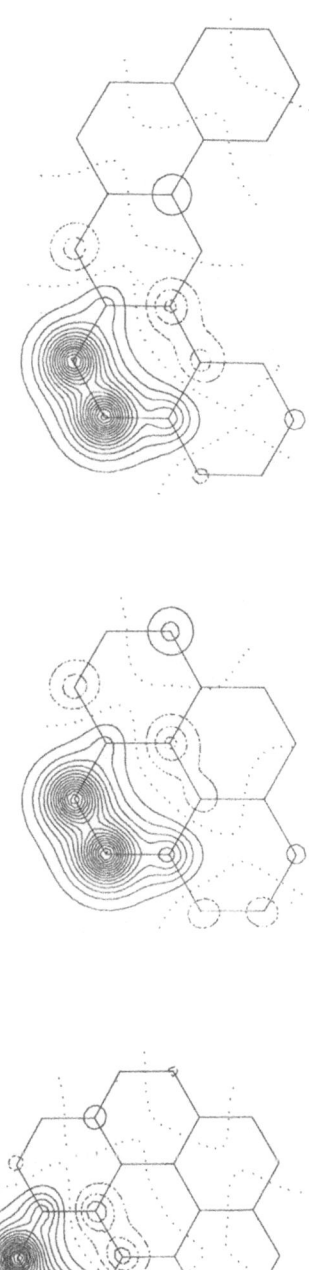

Fig. 19. $\pi\ell 23$ orbitals on branches of two non-joint atoms

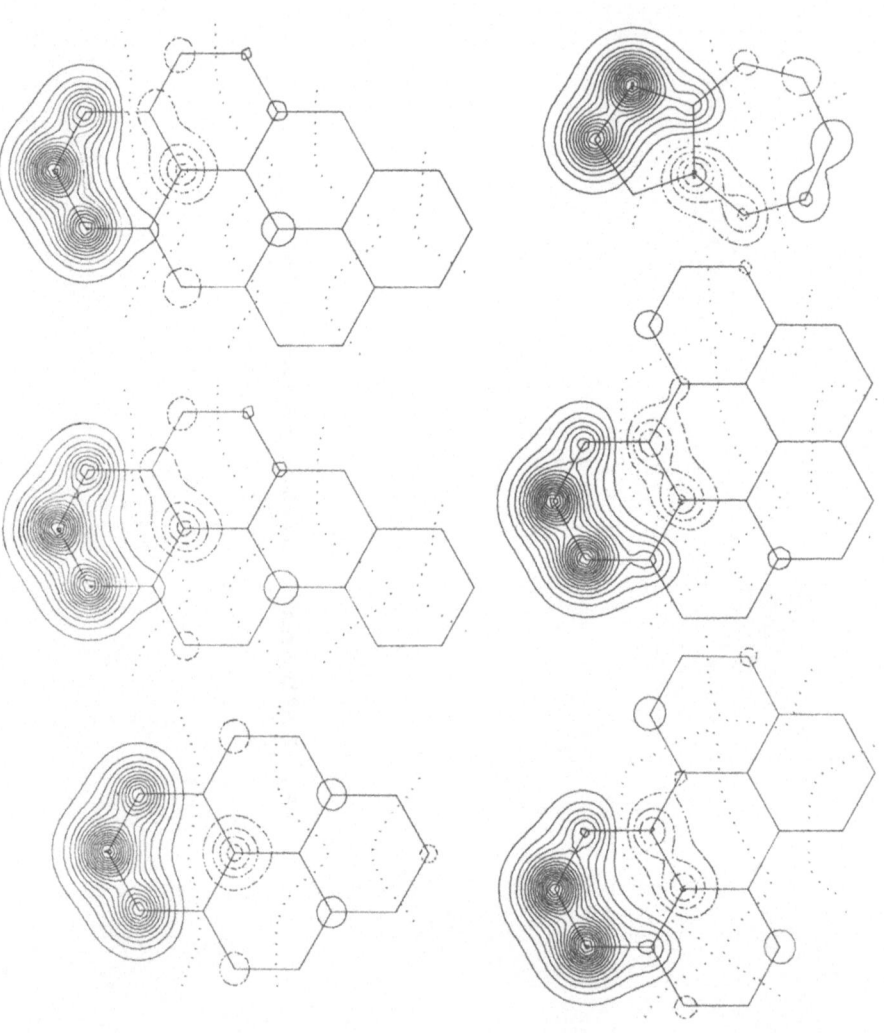

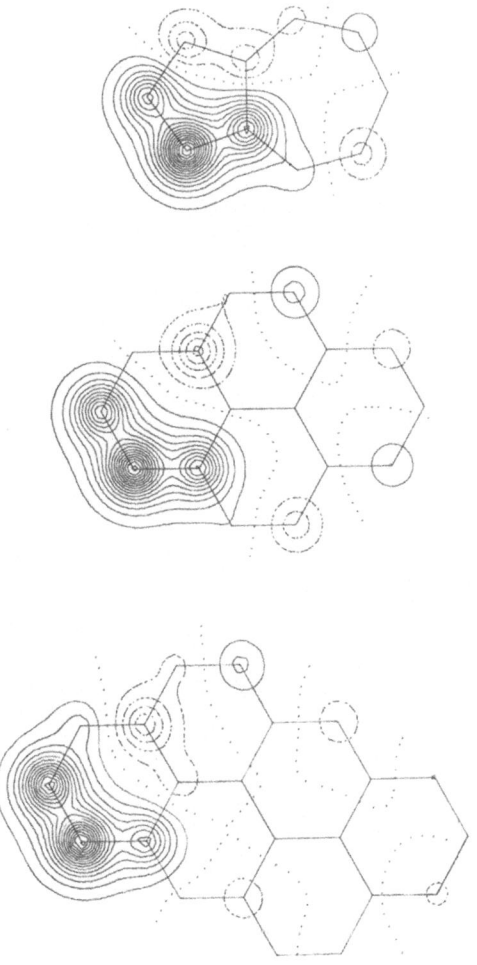

Fig. 20. $\pi\ell 3$-$\pi\ell 2$-$\pi\ell 3$ transition on branches of three non-joint atoms.

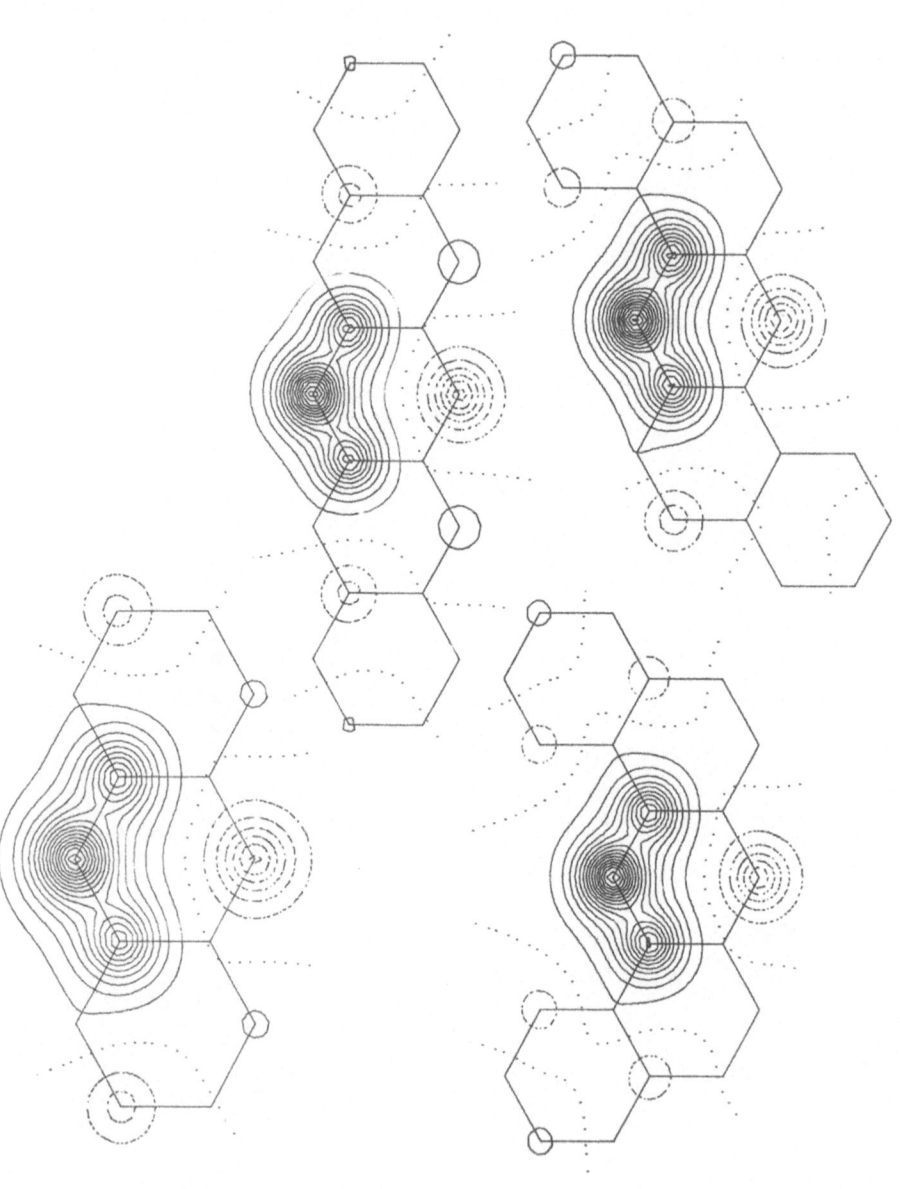

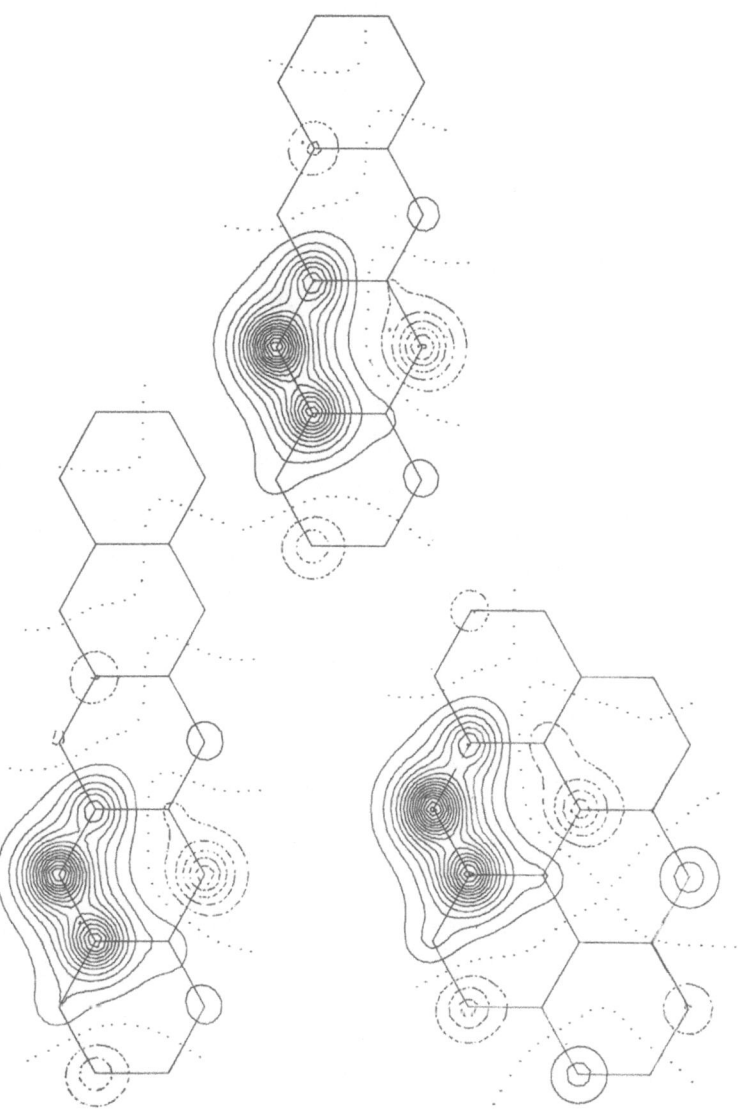

Fig. 21. $\pi\varrho3$–$\pi\varrho34$ transition on branches of one nor-joint atom

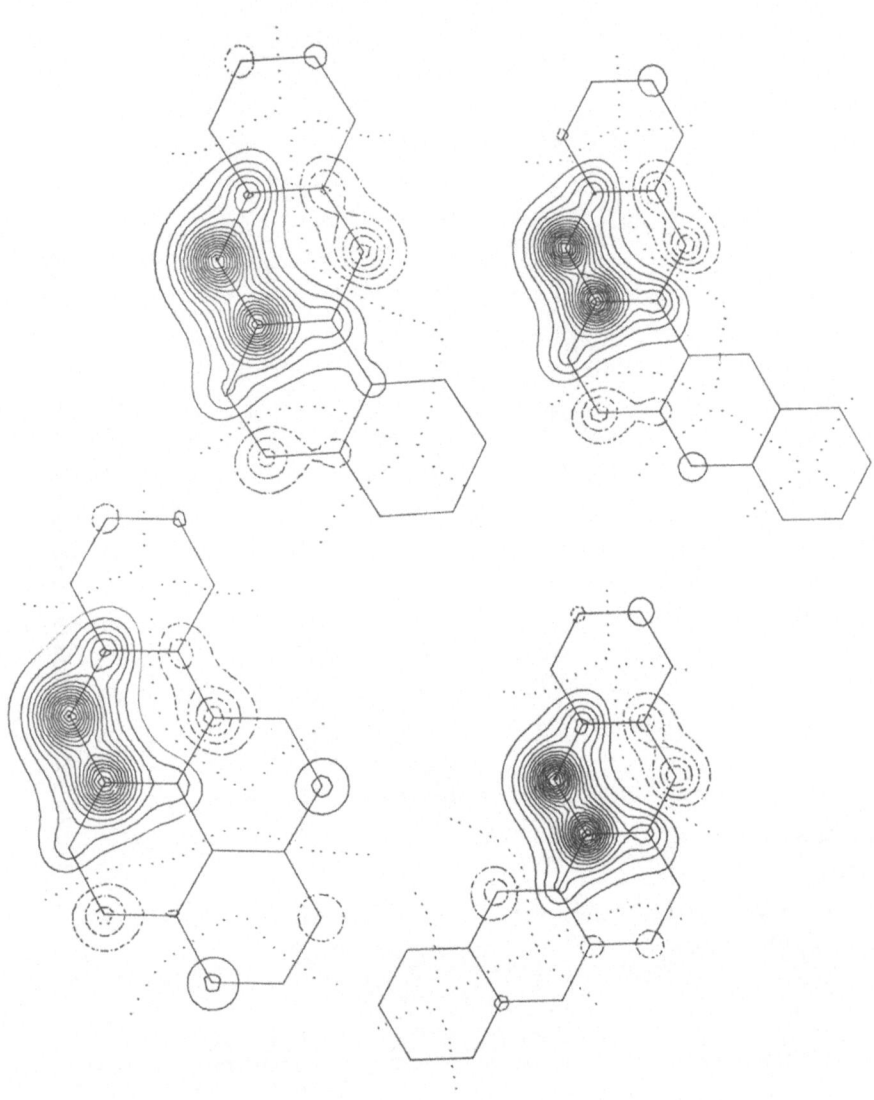

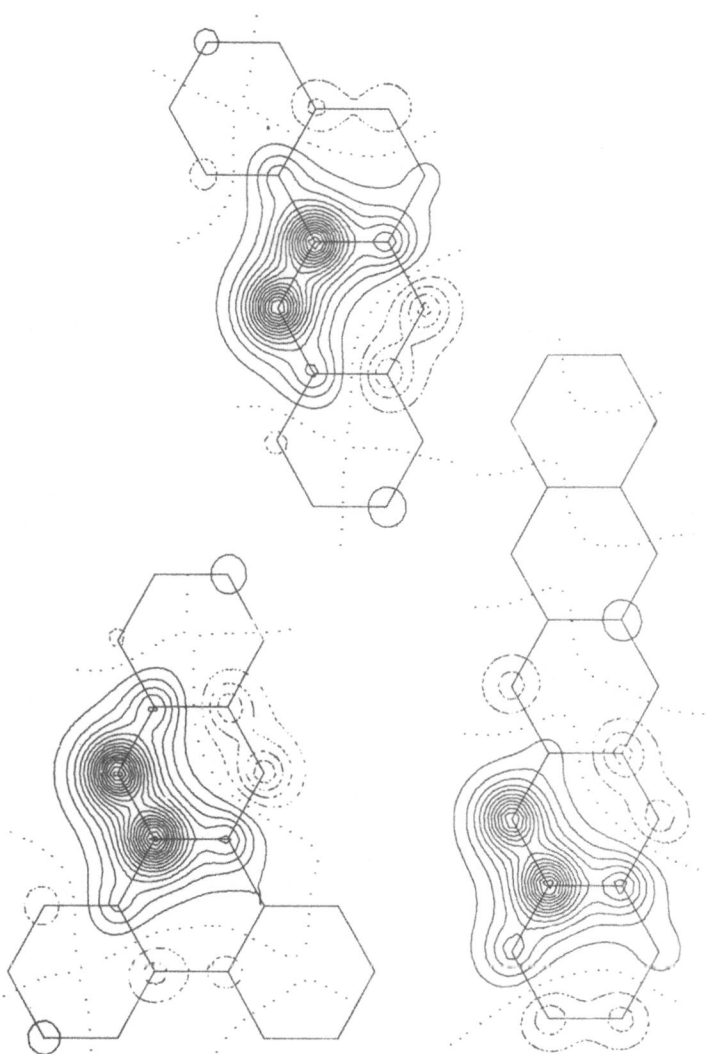

Fig. 22. $\pi\varrho 34$ orbitals on branches of one non-joint atom

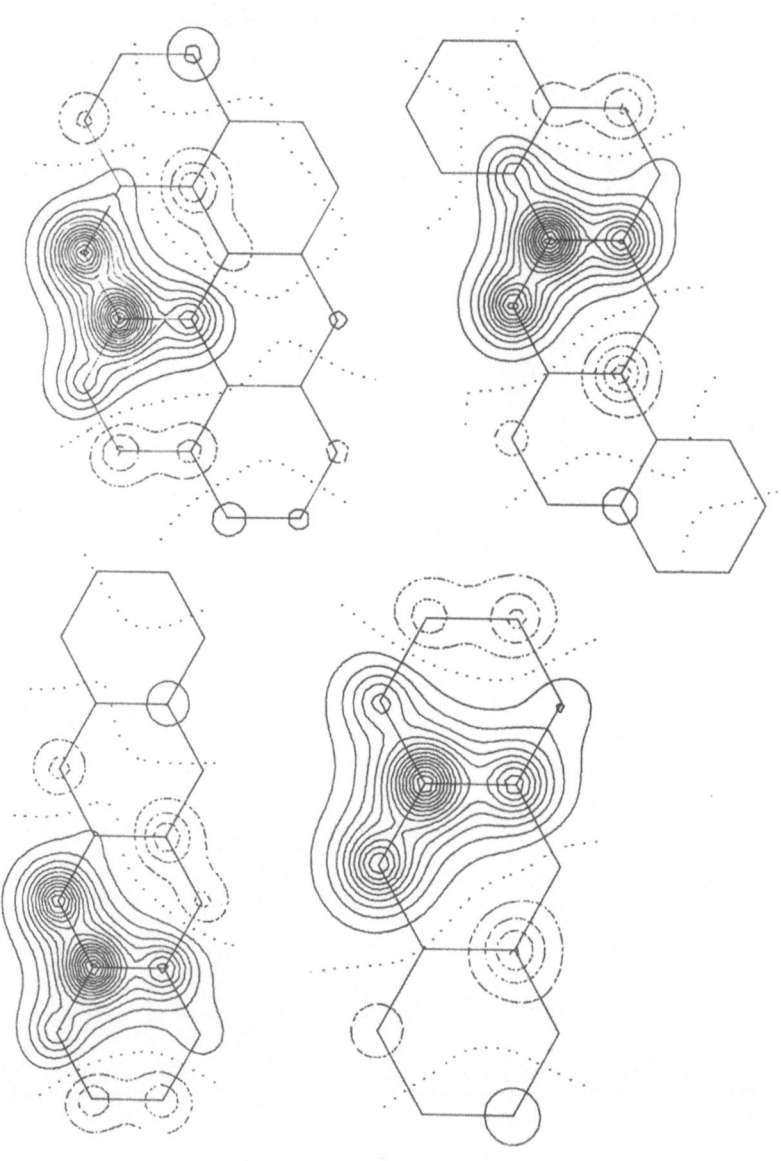

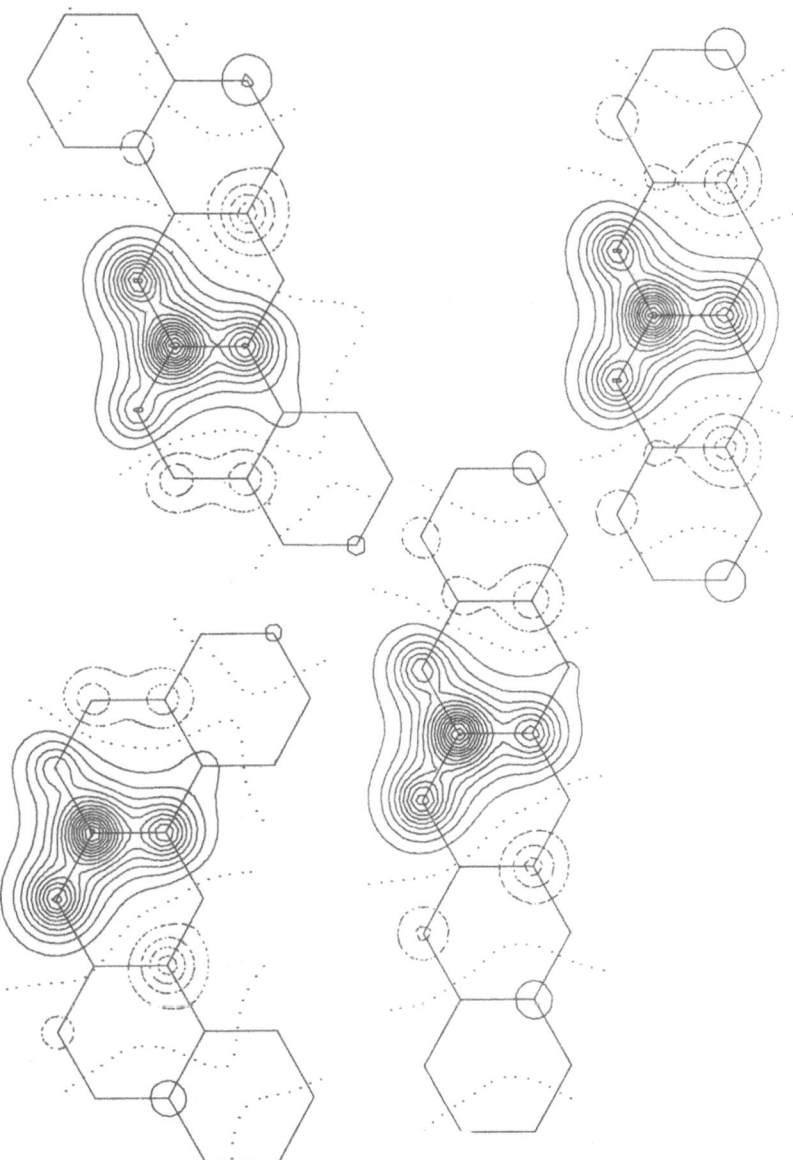

Fig. 23. $\pi\ell 34$–$\pi\ell 4$ transition on branches of one non-joint atom

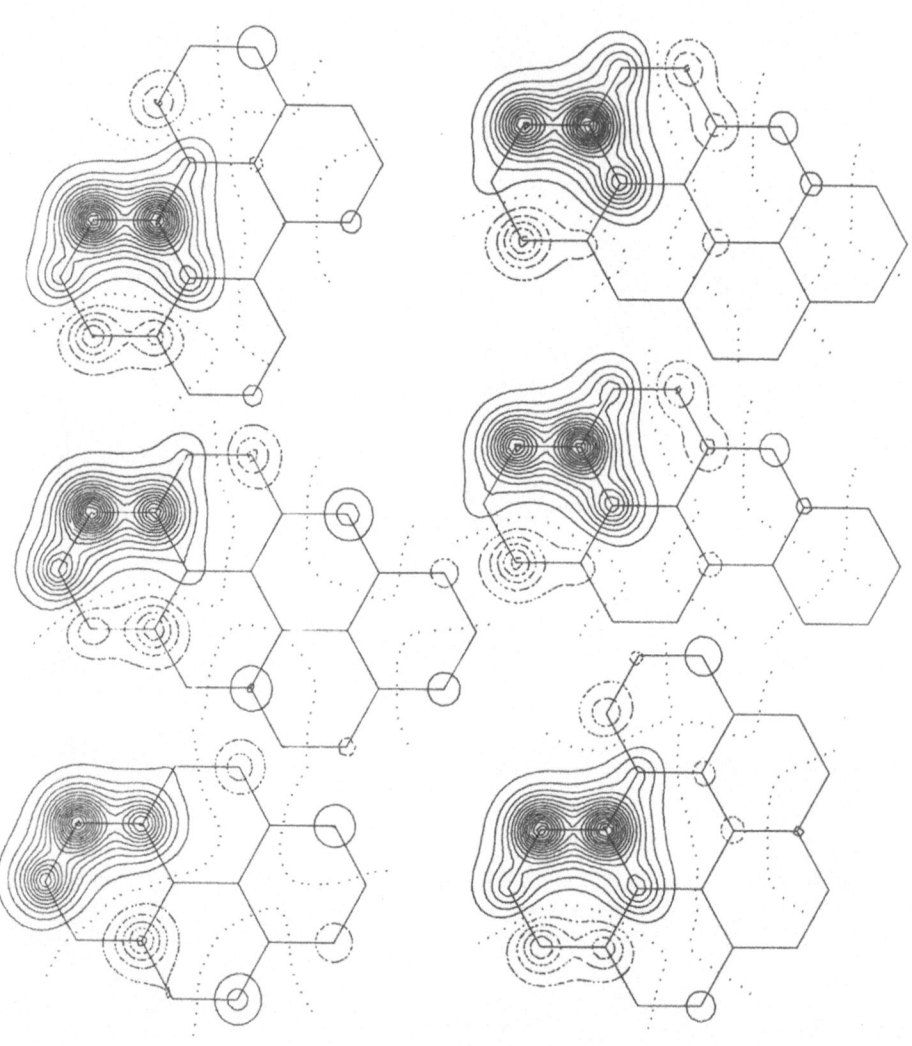

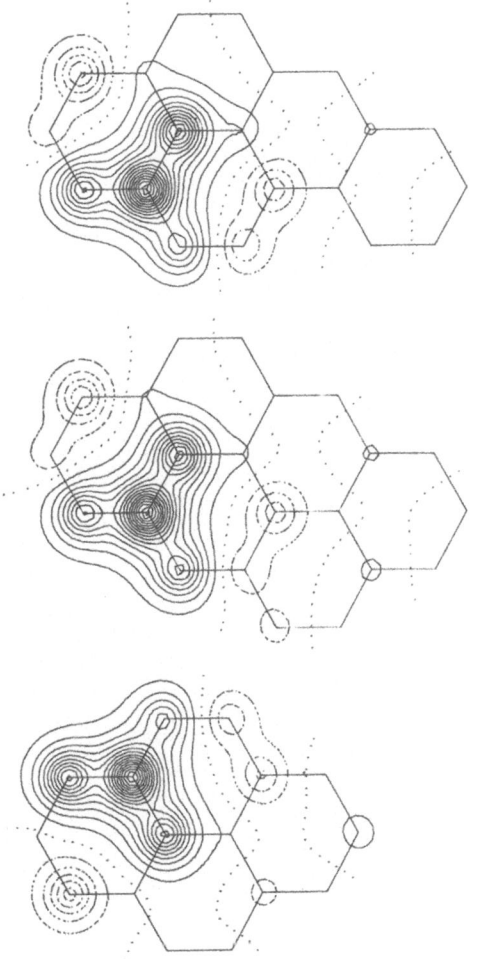

Fig. 24. $\pi\varrho3-\pi\varrho34-\pi\varrho4-\pi\varrho42'$ transition in pericondensed molecules

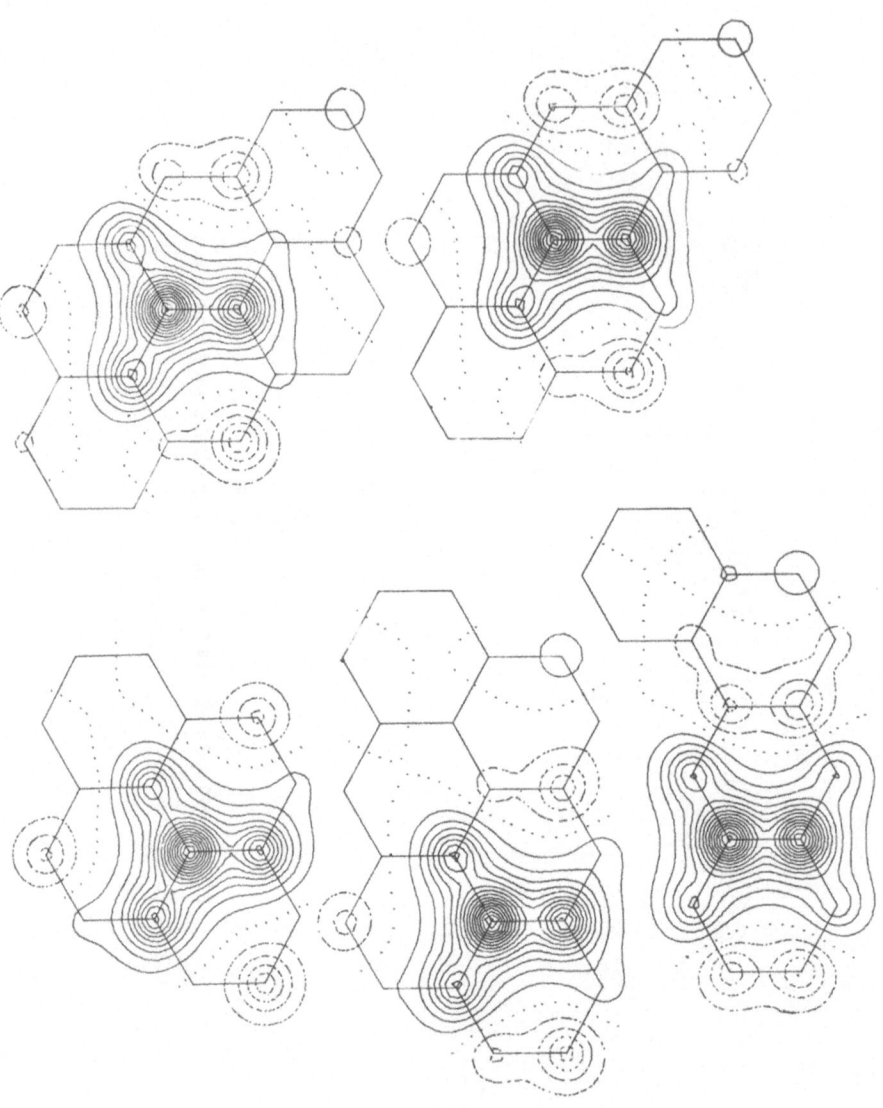

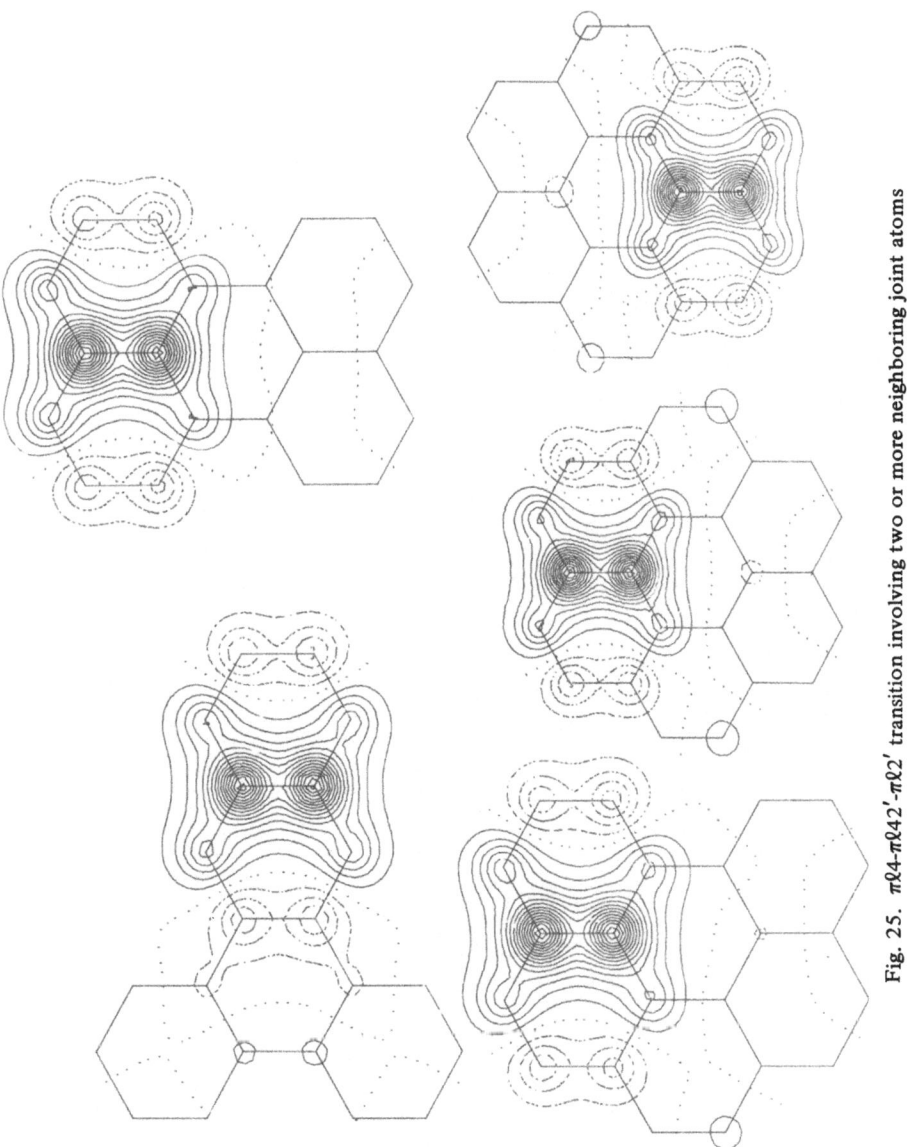

Fig. 25. $\pi\ell 4$-$\pi\ell 42'$-$\pi\ell 2'$ transition involving two or more neighboring joint atoms

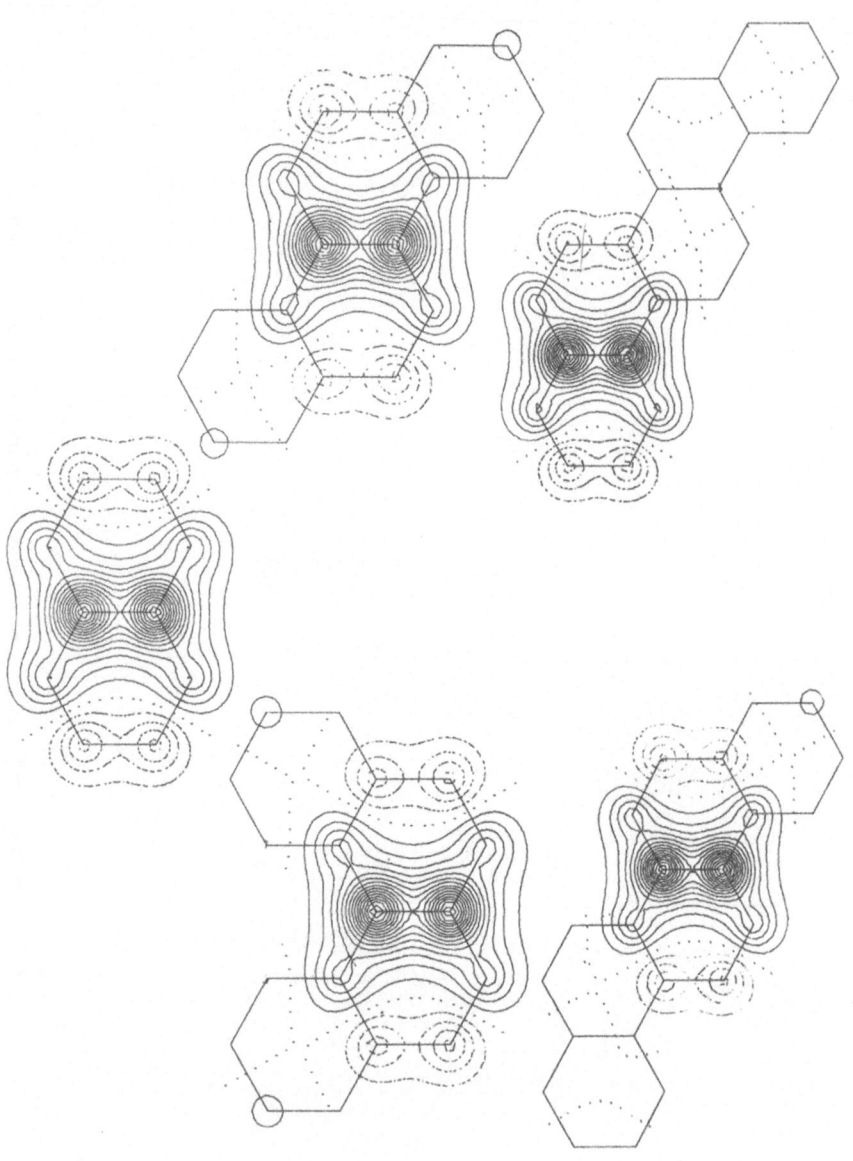

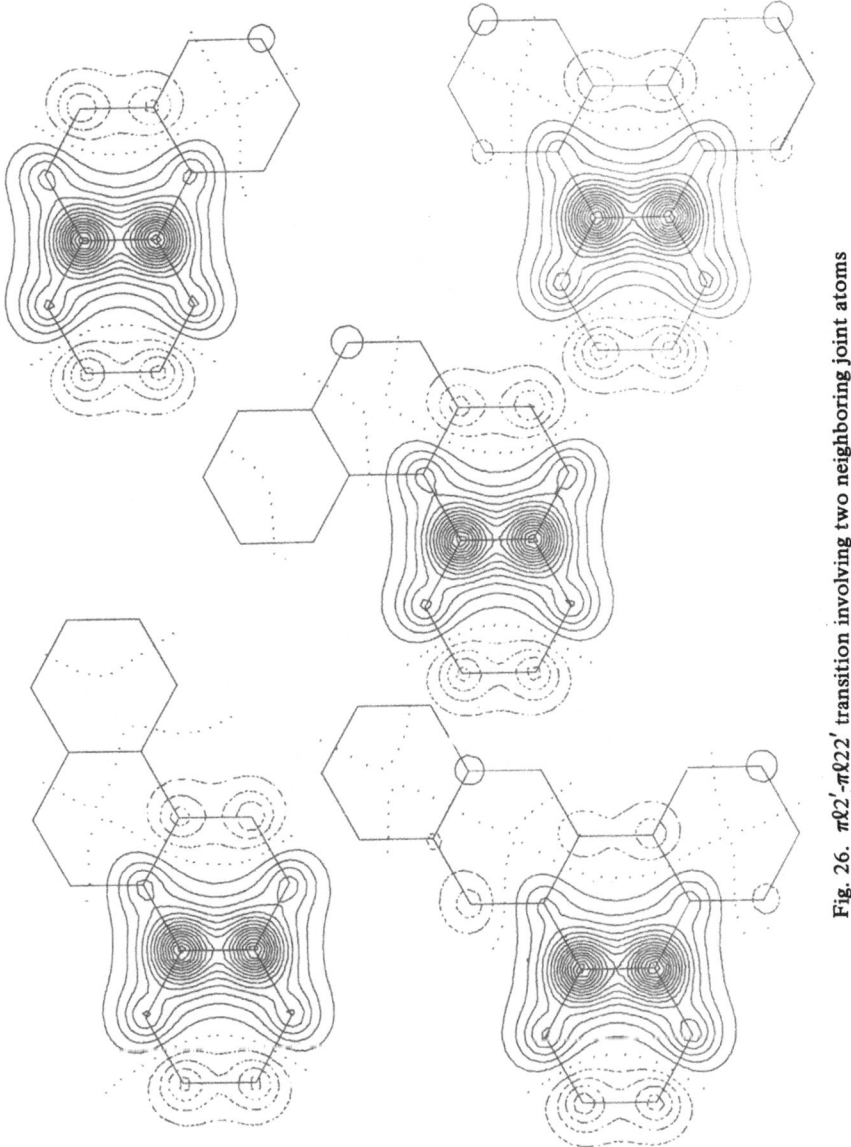

Fig. 26. $\pi\ell 2'-\pi\ell 22'$ transition involving two neighboring joint atoms

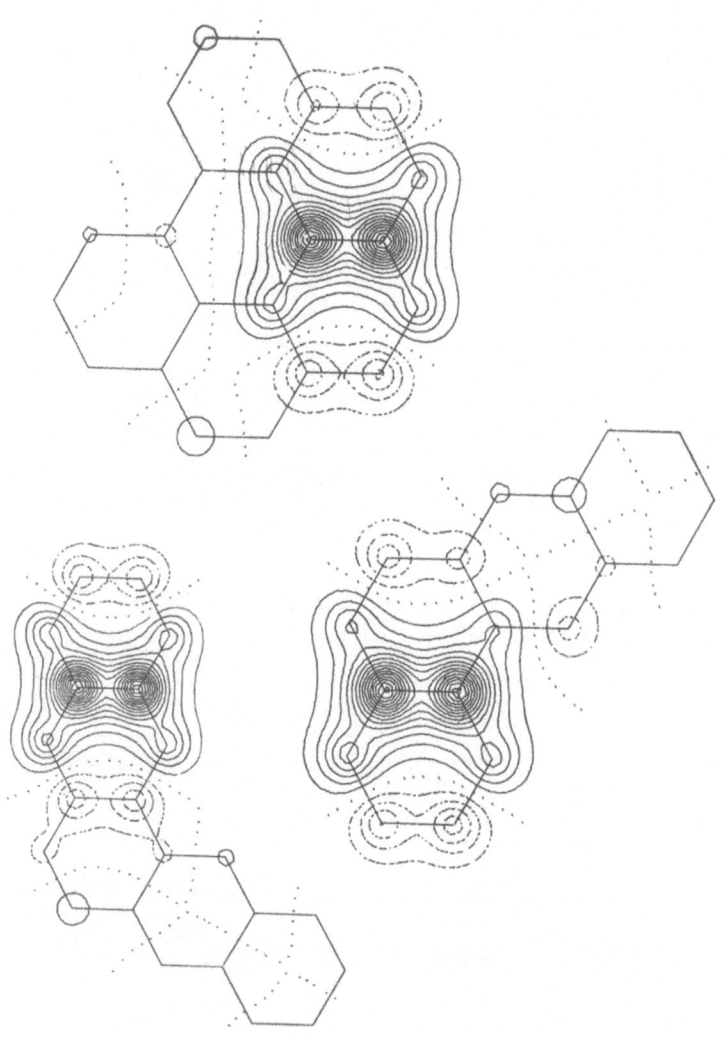

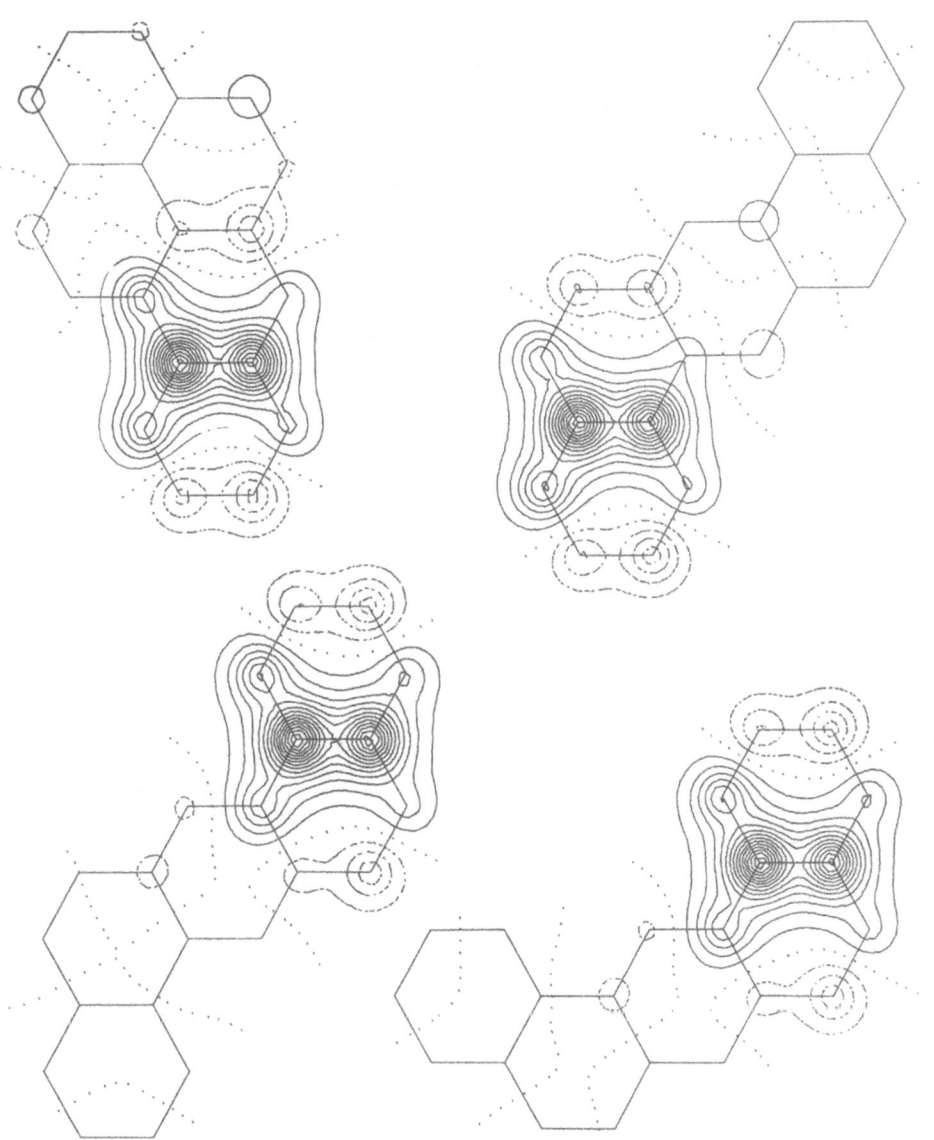

Fig. 27. $\pi\ell 2'$-$\pi\ell 2'4$ transition involving two neighboring joint atoms

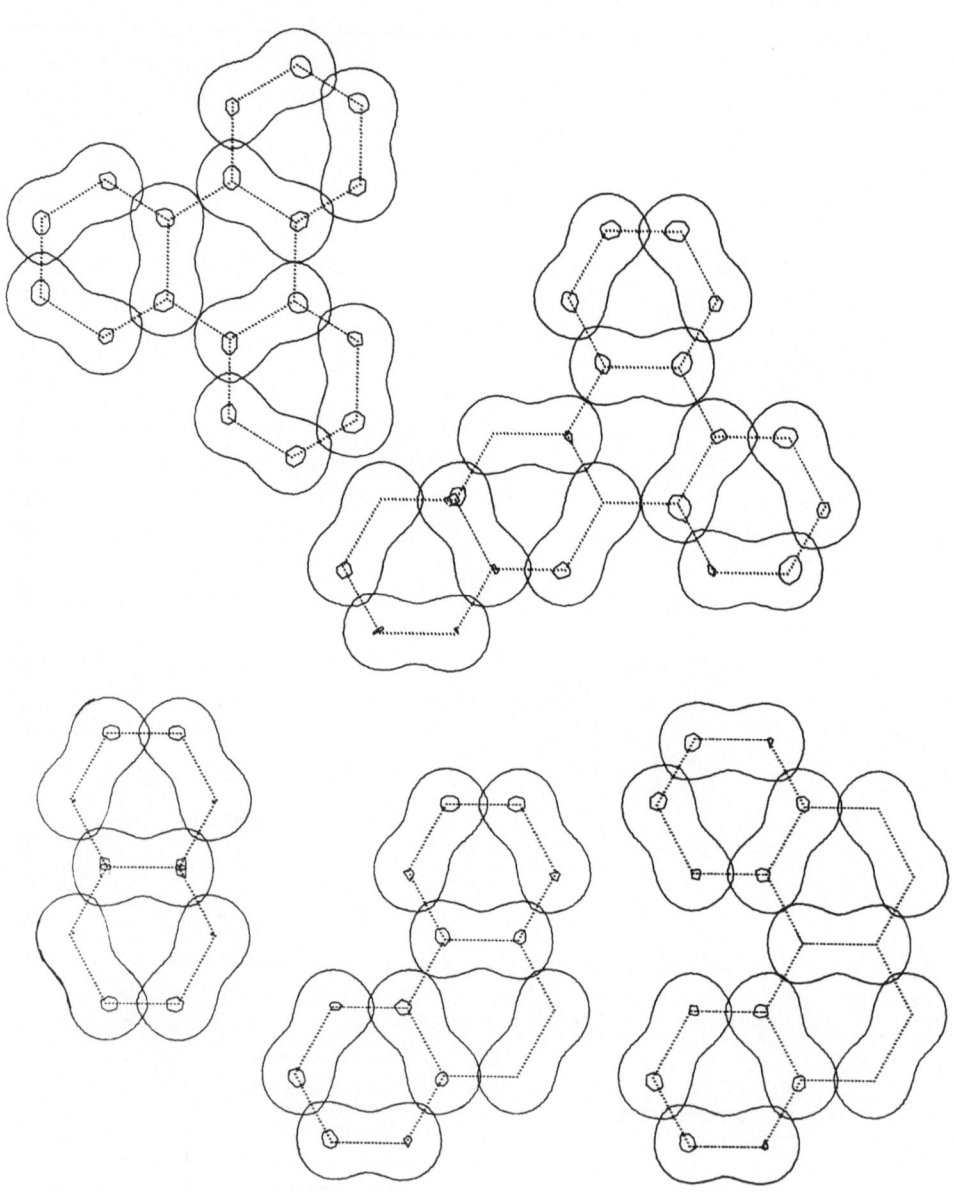

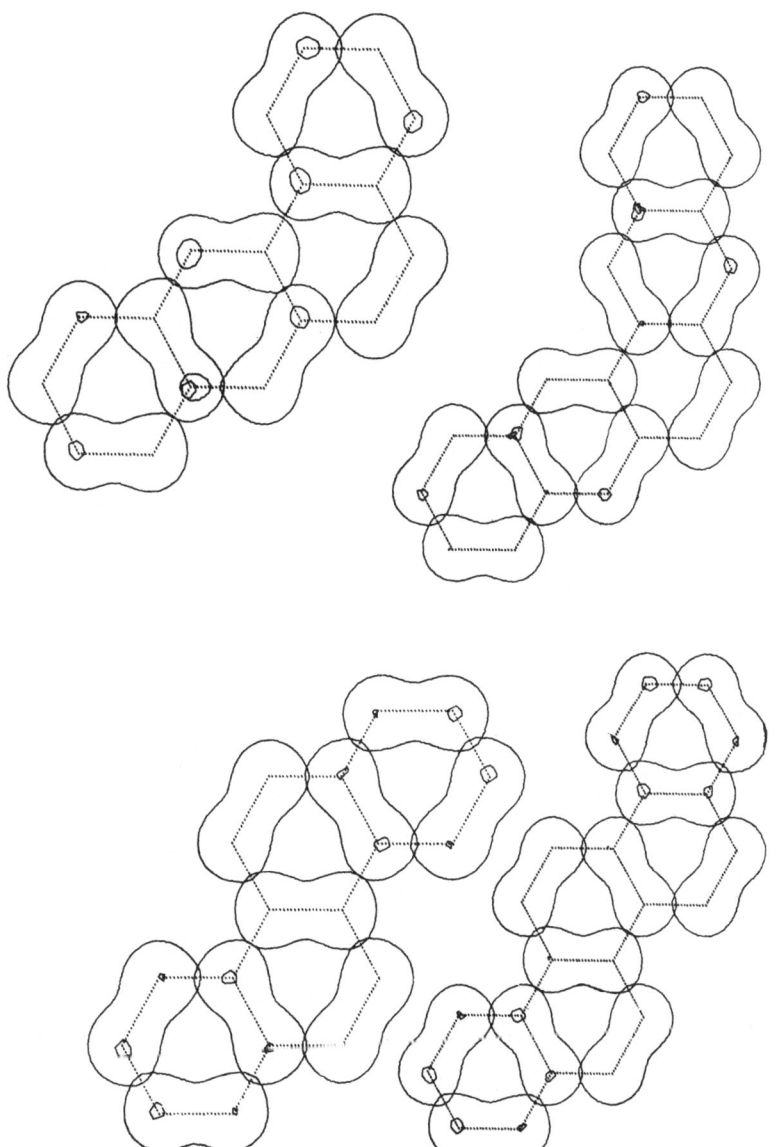

Fig. 28. Localized $\pi$-MO's in catacondensed molecules for which the LMO's correspond to Kekulé structures. Fifth strongest contour is shown

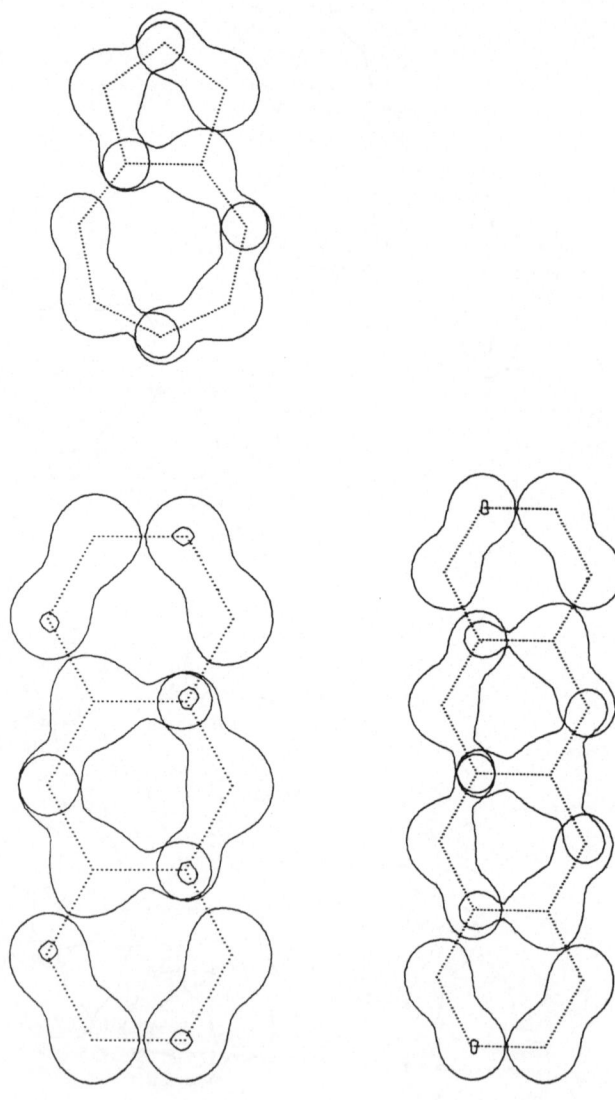

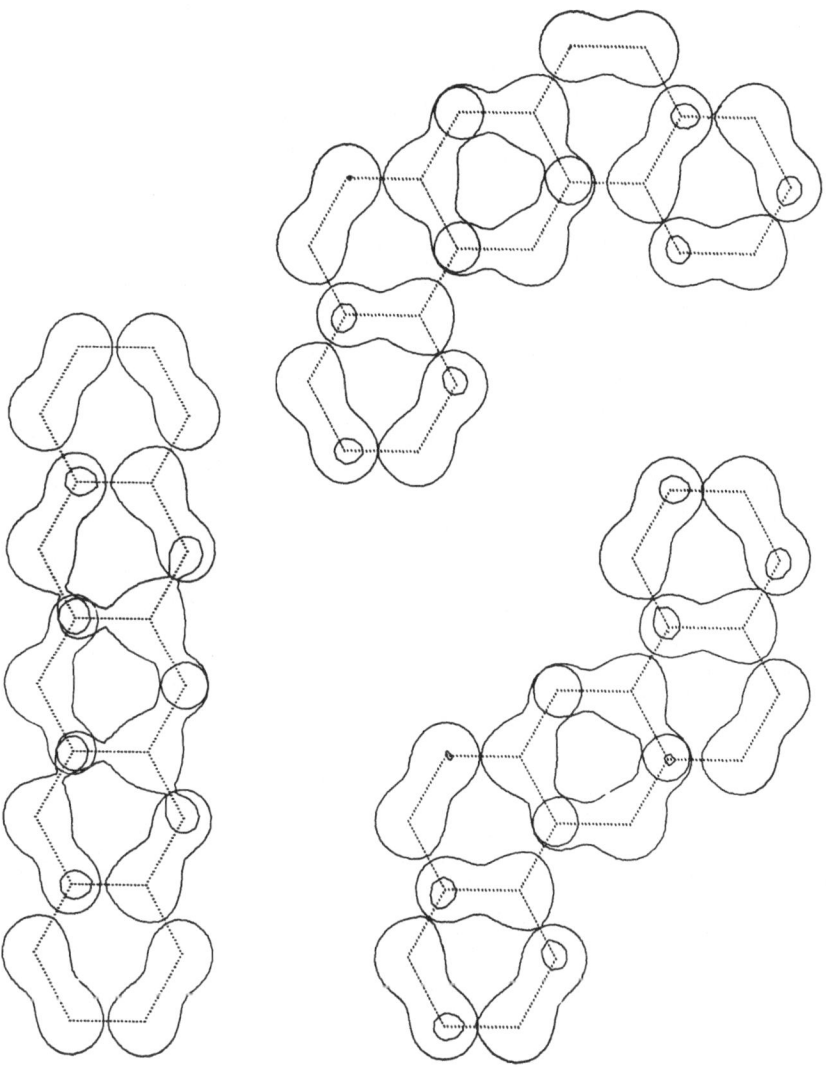

Fig. 29. Localized $\pi$-MO's in catacondensed molecules for which the LMO's do *not* correspond to Kekulé structures. Fifth strongest contour is shown

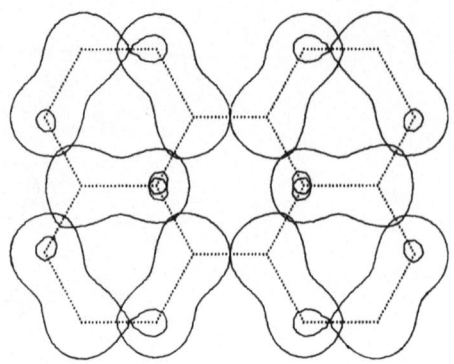

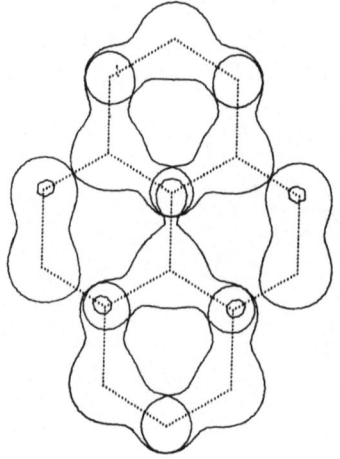

Fig. 30. Localized $\pi$-MO's in

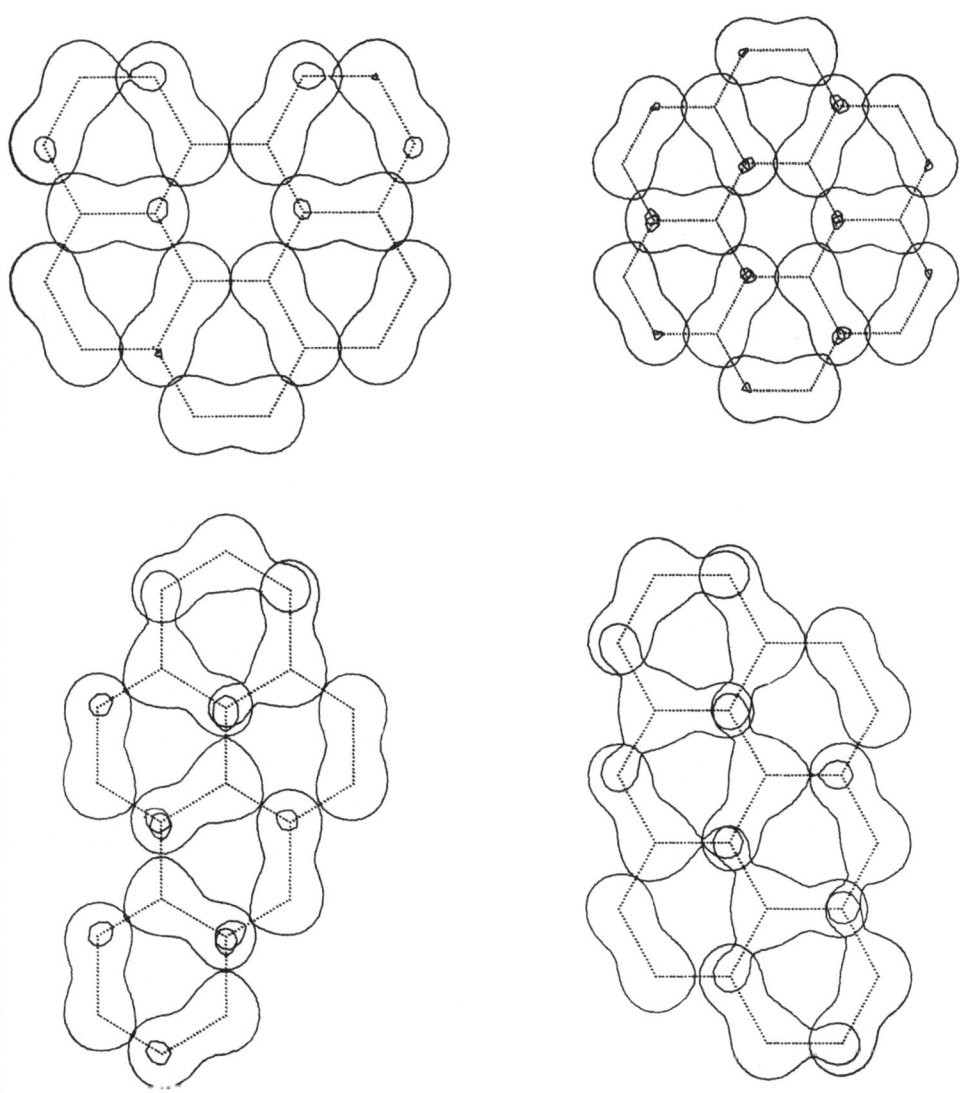

pericondensed molecules. Fifth strongest contour is shown

**Graebe, C.: Geschichte der organischen Chemie**
Erster Band einer Darstellung
in zwei Monographien
(4) X, 406 Seiten. 1920
Reprint 1971
Geb. DM 56,–
Subskriptionspreis bis
30.9.1971:
Geb. DM 48,–

**Walden P.: Geschichte der organischen Chemie seit 1880**
(4) XIV, 946 Seiten. 1941
Reprint 1971
Geb. DM 96,–
Subskriptionspreis bis
30.9.1971:
Geb. DM 84,–

**Grundmann, C., and P. Grünanger: The Nitrile Oxides**
Versatile Tools of
Theoretical and Preparative
Chemistry
1 fig. VIII, 242 pages. 1971
(Organische Chemie in
Einzeldarstellungen, Bd. 13)
Cloth DM 98,–

**Hart, F. L. and H. J. Fisher: Modern Food Analysis**
40 fig. XIV, 530 pages. 1971
Cloth DM 117,10

**Hartmann, H.:**
**Die chemische Bindung**
Drei Vorlesungen für
Chemiker
3. Aufl. 61 Abb.
VI, 109 Seiten. 1971
DM 12,80

**Natural and Synthetic High Polymers**
202 fig. X, 309 pages. 1971
(NMR Basic Principles and
Progress – Grundlagen und
Fortschritte, Vol. 4)
Cloth DM 64,–

**Residue Reviews**

**Vol. 35:** 4 fig. VII, 156 pages
1971. Cloth DM 52,–

**Contents:** Hygienic normalization of pesticide residues and their tolerance in food stuffs in the U.S.S.R. – Materials in contact with foodstuffs: Technical and sanitary grounds in view of a general and specific legislation. – Pesticide legislation and residue problems in Portugal. – Pesticide regulation in South Africa. – Interaction between herbicides and soil microorganisms. – Pesticide and growth regulator residues in pineapple. – Pesticides, pesticide residues, tolerances, and the law (USA).

**Springer-Verlag**
**Berlin**
**Heidelberg**
**New York**
München · London
Paris · Tokyo · Sydney

In kritischen Übersichten werden in dieser Reihe Stand und Entwicklung aktueller chemischer Forschungsgebiete beschrieben. Sie wendet sich an alle Chemiker in Forschung und Industrie, die am Fortschritt ihrer Wissenschaft teilhaben wollen.

In der Regel werden nur Beiträge veröffentlicht, die ausdrücklich angefordert worden sind. Schriftleitung und Herausgeber sind aber für ergänzende Anregungen und Hinweise jederzeit dankbar. Manuskripte können in den „Fortschritten der chemichen Forschung" in Deutsch oder Englisch veröffentlicht werden.

Jeder Band der Reihe ist einzeln käuflich.

This series presents critical reviews of the present position and future trends in modern chemical research. It is addressed to all research and industrial chemists who wish to keep abreast of advances in their subject.

As a rule, contributions are specially commissioned. The editors and publishers will, however, always be pleased to receive suggestions and supplementary information. Papers are accepted for "Topics in Current Chemistry" in either German or English.

Any volume of the series may be purchased separately.